AF355973

Deep Power Vision Technology and Intelligent Vision Sensors

Deep Power Vision Technology and Intelligent Vision Sensors

Editors

Ke Zhang
Yincheng Qi

Basel • Beijing • Wuhan • Barcelona • Belgrade • Novi Sad • Cluj • Manchester

Editors
Ke Zhang
Department of Electronic
Communication Engineering
North China Electric Power
University
Baoding
China

Yincheng Qi
Department of Electronic
Communication Engineering
North China Electric Power
University
Baoding
China

Editorial Office
MDPI
St. Alban-Anlage 66
4052 Basel, Switzerland

This is a reprint of articles from the Special Issue published online in the open access journal *Sensors* (ISSN 1424-8220) (available at: www.mdpi.com/journal/sensors/special_issues/DPVTIVS).

For citation purposes, cite each article independently as indicated on the article page online and as indicated below:

Lastname, A.A.; Lastname, B.B. Article Title. *Journal Name* **Year**, *Volume Number*, Page Range.

ISBN 978-3-7258-0016-2 (Hbk)
ISBN 978-3-7258-0015-5 (PDF)
doi.org/10.3390/books978-3-7258-0015-5

Contents

Preface

This reprint is about deep power vision technology and intelligent vision sensors, which is the application of deep learning-based computer vision technology in power systems and is an important component of power artificial intelligence technology. This work aims to provide some up-to-date solutions to process inspection images of the power system obtained with intelligent vision sensors via deep learning-based computer vision algorithms. We hope this work will provide some inspiration to the researchers or engineers from academic or industrial backgrounds and further boost the fundamental and practical research in the direction of deep power vision technology. Meanwhile, we thank all authors who submitted their manuscripts as well as the reviewers for their invaluable input and comments. We also thank the editors of *Sensors* for their support.

Ke Zhang and Yincheng Qi

Editors

 *sensors*

Editorial

Deep Power Vision Technology and Intelligent Vision Sensors

Ke Zhang * and Yincheng Qi

Department of Electronic & Communication Engineering, School of Electrical and Electronic Engineering, North China Electric Power University, 619 Yonghuabei Dajie, Baoding 071000, China; qiych@ncepu.edu.cn
* Correspondence: zhangkeit@ncepu.edu.cn

Citation: Zhang, K.; Qi, Y. Deep Power Vision Technology and Intelligent Vision Sensors. *Sensors* **2023**, *23*, 9626. https://doi.org/10.3390/s23249626

Received: 22 November 2023
Accepted: 23 November 2023
Published: 5 December 2023

1. Introduction

With the rapid development of the power system and the increasing burden of its inspection, more attention has been paid to the automatic inspection technologies based on deep power vision technology and intelligent vision sensors. Deep power vision technology is aiming at processing and analyzing the inspection images and videos obtained through vision sensors on unmanned aerial vehicles or robots with deep learning-based computer vision algorithms. In the latest research, deep power vision technology has been widely used in the scene of processing the goals and defects of power plants, transmission lines, substations, and distribution lines in electric power systems.

2. Overview of Contribution

This Special Issue aims to provide some up-to-date solutions to the problems of inspection in the power system and offer helpful reference for further research of deep power vision technology and intelligent vision sensors. It includes ten papers covering the tasks of detection for the inspection of transmission lines and substation [1–7], classification related to the requirements of inspection [8,9], and image defogging for transmission lines [10].

To achieve a better performance in the detection tasks of the transmission line, the following methods were proposed. Wang et al. [1] proposed a fitting detection method based on multi-scale geometric transformation and an attention-masking mechanism, which demonstrated its effectiveness in improving the detection accuracy of transmission line fittings. In order to solve the problem of background interference and overlap caused by the axis-aligned bounding boxes in the tilting insulator detection tasks, Zhao et al. [2] designed a normal orientation detection method incorporating the angle regression and a priori constraints. In terms of restraining the negative impact of the large-scale gap of the fittings in the transmission line inspection, Zhao et al. [3] developed an optimized method based on contextual information enhancement and joint heterogeneous representation. Han et al. [4] were concerned with the shortcomings of IoU as well as the sensitivity of small targets to the model regression accuracy and proposed an improved YOLOX to solve the problem of low accuracy of insulator defect detection. Zhai et al. [5] introduced a multi-geometric reasoning network to accurately detect insulator geometric defects based on aerial images with complex backgrounds and different scales which significantly improved the detection accuracy of multiple insulator defects using aerial images. Xin et al. [6] combined the defogging algorithm with a two-stage detection model in order to accomplish the accurate detection of the insulator umbrella disc shedding in foggy weather. And for better inspection in a substation, Li et al. [7] presented a two-level defect detection model for alleviating the adverse effect of the complex background of substation in the infrared images.

In order to explore the solution of classification problems related to the requirements of inspection, two research articles were included in this issue. Since there are always limited defect data existing in the power system, Wang et al. [8] incorporated a semantic information fusion method based on matrix decomposition and a spatial attention mechanism to improve the classification accuracy for unseen images. For accurate identification of the bolt defect, Liu et al. [9] proposed a bolt defect identification method in which an attention mechanism and wide residual networks were combined.

This Issue also contains a method of image defogging under the scene of transmission line inspection. In terms of the fuzziness and the concealment problems in inspection images caused by fogs, Zai et al. [10] created the UAV-HAZE dataset for power assessment of unmanned aerial vehicles and presented a dual attention level feature fusion multi-patch hierarchical network for single-image defogging.

3. Conclusions

In conclusion, though a wide range of solutions to a variety of problems in the inspection of power system are presented in this Special Issue, the investigation of deep power vision technology and intelligent vision sensors still has a long way to go. We hope this Special Issue will provide some inspiration to researchers or engineers from academic or industrial backgrounds and further boost the fundamental and practical research in this direction.

Author Contributions: K.Z. and Y.Q. contributed equally. All authors have read and agreed to the published version of the manuscript.

Funding: This work was supported in part by the National Natural Science Foundation of China (NSFC) under Grant 62076093, Grant 62206095, and Grant 61871182; and in part by the Fundamental Research Funds for the Central Universities under Grant 2023JG002, Grant 2022MS078 and Grant 2023JC006.

Acknowledgments: The Guest Editors would like to thank all authors who submitted their manuscripts to this Special Issue and the reviewers for their invaluable input and comments. We would also like to thank the editors of Sensors for their support.

Conflicts of Interest: The authors declare no conflict of interest.

References

1. Wang, N.; Zhang, K.; Zhu, J.; Zhao, L.; Huang, Z.; Wen, X.; Zhang, Y.; Lou, W. Fittings Detection Method Based on Multi-Scale Geometric Transformation and Attention-Masking Mechanism. *Sensors* **2023**, *23*, 4923. [CrossRef] [PubMed]
2. Zhao, J.; Liu, L.; Chen, Z.; Ji, Y.; Feng, H. A New Orientation Detection Method for Tilting Insulators Incorporating Angle Regression and Priori Constraints. *Sensors* **2022**, *22*, 9773. [CrossRef] [PubMed]
3. Zhao, L.; Liu, C.; Qu, H. Transmission Line Object Detection Method Based on Contextual Information Enhancement and Joint Heterogeneous Representation. *Sensors* **2022**, *22*, 6855. [CrossRef] [PubMed]
4. Han, G.; Li, T.; Li, Q.; Zhao, F.; Zhang, M.; Wang, R.; Yuan, Q.; Liu, K.; Qin, L. Improved Algorithm for Insulator and Its Defect Detection Based on YOLOX. *Sensors* **2022**, *22*, 6186. [CrossRef] [PubMed]
5. Zhai, Y.; Hu, Z.; Wang, Q.; Yang, Q.; Yang, K. Multi-Geometric Reasoning Network for Insulator Defect Detection of Electric Transmission Lines. *Sensors* **2022**, *22*, 6102. [CrossRef] [PubMed]
6. Xin, R.; Chen, X.; Wu, J.; Yang, K.; Wang, X.; Zhai, Y. Insulator Umbrella Disc Shedding Detection in Foggy Weather. *Sensors* **2022**, *22*, 4871. [CrossRef] [PubMed]
7. Li, B.; Wang, T.; Hu, Z.; Yuan, C.; Zhai, Y. Two-Level Model for Detecting Substation Defects from Infrared Images. *Sensors* **2022**, *22*, 6861. [CrossRef] [PubMed]
8. Wang, Y.; Feng, L.; Song, X.; Xu, D.; Zhai, Y. Zero-Shot Image Classification Method Based on Attention Mechanism and Semantic Information Fusion. *Sensors* **2023**, *23*, 2311. [CrossRef] [PubMed]
9. Liu, L.; Zhao, J.; Chen, Z.; Zhao, B.; Ji, Y. A New Bolt Defect Identification Method Incorporating Attention Mechanism and Wide Residual Networks. *Sensors* **2022**, *22*, 7416. [CrossRef] [PubMed]
10. Zai, W.; Yan, L. Multi-Patch Hierarchical Transmission Channel Image Dehazing Network Based on Dual Attention Level Feature Fusion. *Sensors* **2023**, *23*, 7026. [CrossRef] [PubMed]

 *sensors*

Article

Fittings Detection Method Based on Multi-Scale Geometric Transformation and Attention-Masking Mechanism

Ning Wang [1,*], Ke Zhang [2,*], Jinwei Zhu [1], Liuqi Zhao [1], Zhenlin Huang [1], Xing Wen [1], Yuheng Zhang [1] and Wenshuo Lou [2]

1   Operation and Maintenance Center of Information and Communication, CSG EHV Power Transmission Company, Guangzhou 510000, China; zhujinwei@ehv.csg.cn (J.Z.); zhaoliuqi@ehv.csg.cn (L.Z.); huangzhenlin@ehv.csg.cn (Z.H.); wenxing@ehv.csg.cn (X.W.); zhangyuheng@ehv.csg.cn (Y.Z.)
2   Department of Electronic and Communication Engineering, North China Electric Power University, Baoding 071003, China; lws_sure@163.com
*   Correspondence: wangning1@ehv.csg.cn (N.W.); zhangkeit@ncepu.edu.cn (K.Z.)

Abstract: Overhead transmission lines are important lifelines in power systems, and the research and application of their intelligent patrol technology is one of the key technologies for building smart grids. The main reason for the low detection performance of fittings is the wide range of some fittings' scale and large geometric changes. In this paper, we propose a fittings detection method based on multi-scale geometric transformation and attention-masking mechanism. Firstly, we design a multi-view geometric transformation enhancement strategy, which models geometric transformation as a combination of multiple homomorphic images to obtain image features from multiple views. Then, we introduce an efficient multiscale feature fusion method to improve the detection performance of the model for targets with different scales. Finally, we introduce an attention-masking mechanism to reduce the computational burden of model-learning multiscale features, thereby further improving model performance. In this paper, experiments have been conducted on different datasets, and the experimental results show that the proposed method greatly improves the detection accuracy of transmission line fittings.

Keywords: geometric transformation; fittings; object detection; transformer

Citation: Wang, N.; Zhang, K.; Zhu, J.; Zhao, L.; Huang, Z.; Wen, X.; Zhang, Y.; Lou, W. Fittings Detection Method Based on Multi-Scale Geometric Transformation and Attention-Masking Mechanism. *Sensors* **2023**, *23*, 4923. https://doi.org/10.3390/s23104923

Academic Editor: Paweł Pławiak

Received: 17 April 2023
Revised: 17 May 2023
Accepted: 18 May 2023
Published: 19 May 2023

1. Introduction

With the development of the economy, the scale of equipment in the power system continues to expand. In order to explore the application prospects and directions of cutting-edge technologies such as artificial intelligence in the field of power, the development of human-machine interaction intelligent systems with reasoning, perception, self training, and learning abilities has become increasingly important research in the field of power [1].

Currently, the length of the power system's overhead transmission lines has reached 992,000 km and still maintains an annual growth rate of about 5%. Overhead transmission lines are distributed in vast outdoor areas with complex geographical environments, and the traditional manual inspection mode is inefficient [2,3]. In response to the increasingly prominent contradiction between the number of transmission professionals and the continuous growth of equipment scale, the power system promoted the application of unmanned aerial vehicle (UAV) patrol inspection, significantly improving the efficiency of transmission line patrol inspection [4–6]. Figure 1 shows patrol inspection images of a transmission line taken by the UAV.

The development of artificial intelligence technology, represented by deep learning, provides theoretical support for the transformation of the overhead transmission line inspection mode from manual inspection to intelligent inspection based on UAV [7]. Object detection is a fundamental task in the field of computer vision. Currently, popular object detection methods mainly use convolutional neural networks (CNN) and Transformer architecture to extract and learn image features. The object detection method based on CNN can be divided into two-stage detection models [8–11] based on candidate frame generation and single-stage detection models [12–14] based on regression. In recent years, the Transformer model for computer vision tasks has been studied by many scholars [15]. Carion et al. [16] proposed the DETR model which uses an encode–decode structured Transformer. Given a fixed set of target sequences, the relationship between the targets and the global context of the image can be inferred, and the final prediction set can be output directly and in parallel, avoiding the manual design. Zhu et al. [17] proposed Deformable DETR, in which the attention module only focuses on a portion of key sampling points around the reference point. With $10\times$ less training epochs, Deformable DETR can achieve better performance than DETR. Roh et al. [18] propose Sparse DETR, which helps the model effectively detect targets by selectively updating only some tokens. Experiments have shown that even with only 10% of encoder tokens, the Sparse DETR can achieve better performance. Fang et al. [19] propose to use only Transformer's encoder for target detection, further reducing the weight of the Transformer-based target detection model at the expense of target detection accuracy. Song et al. [20] introduce a computationally efficient Transformer decoder that utilizes multi-scale features and auxiliary techniques to improve detection performance without increasing too much computational load. Wu et al. [21] proposed an image relative position encoding method for two-dimensional images. This method considers the interaction between direction, distance, query, and relative position encoding in the self-attention mechanism, further improving the performance of target detection.

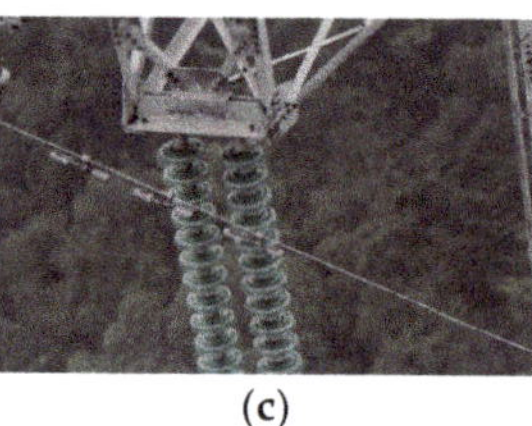

(a)   (b)   (c)

Figure 1. Transmission line images captured by the UAV.

Applying the object detection models that perform well in the field of general object detection to power component detection has become a hot research topic in the current power field [22–25]. Zhao et al. [26] use a CNN model with multiple feature extraction methods to represent the status of insulators, and train support vector machines based on these features to detect the status of insulators. Zhao et al. [27] designed an intelligent monitoring system for hazard sources on transmission lines based on deep learning, which can accurately identify hazard sources and ensure the safe operation of the power system. Zhang et al. [28] propose a high-resolution real-time network HRM-CenterNet, which utilizes iterative aggregation of high-resolution feature fusion methods to gradually fuse high-level and low-level information to improve the detection accuracy of fittings in transmission lines. Zhang et al. [29] first proposed that there is a visual indivisibility problem with bolt defects on transmission lines and that the attributes of bolts, such as whether there are pin holes or gaskets, are visually separable. Therefore, bolt recognition is considered a multi-attribute classification problem, and a multi-label classification method is used to obtain accurate bolt multi-attribute information. Lou et al. [30] introduce the position knowledge and attribute knowledge of bolts into the model for the detection of visually indivisible bolt defects, further improving the detection accuracy of visually indivisible bolt defects.

Although there have been some related studies on transmission line fittings detection in the field of electric power, quite difficult problems remain. The main performance is shown in the following aspects: (1) Due to the variable viewing angles of UAV photography, the shape of some fittings varies greatly under different shooting visions, resulting in poor detection performance of fittings under different viewing angles. As shown in Figure 2, the blue frame is the bag-type suspension clamp, and the red frame is the weight. As can be seen from Figure 2, the appearance of the bag-type suspension clamp and the weight has undergone significant changes under different shooting visions. (2) Figure 3 shows the area ratio of different fittings tags in different transmission line datasets. It can be seen that the scale of different fittings in each dataset varies greatly, which is an important factor affecting detection performance. (3) The UAV edge device is small in size and has limited storage and computational resources, so the detection model cannot be too complex. To address the above issues, this paper proposes a transmission line fittings detection method based on multi-scale geometric transformation and attention-masking mechanism (MGA-DETR). The main contributions of this article are as follows:

1. We have designed a multi-view geometric transformation enhancement strategy that models geometric transformations as a combination of multiple homomorphic images to obtain image features from multiple views. At the same time, this paper introduces an efficient multi-scale feature fusion method to improve the detection performance of transmission line fittings from different perspectives and scales.
2. We introduced an attention-masking mechanism to reduce the computational burden of model-learning multiscale features, thereby further improving the detection speed of the model without affecting its detection accuracy.
3. We conducted experiments on three different sets of transmission line fittings detection data, and the experimental results show that the method proposed in this paper can effectively improve the detection accuracy of different scale fittings from different perspectives.

Figure 2. Transmission line images from different shooting angles.

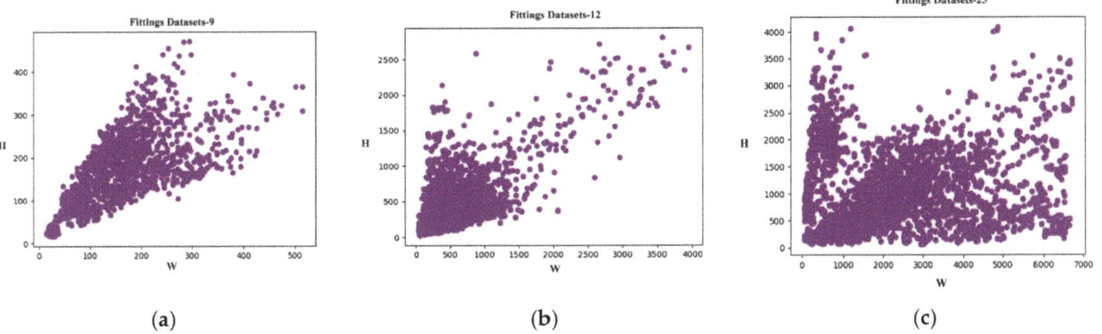

Figure 3. Scale distribution of fittings in different transmission line datasets.

The rest of the paper is organized as follows: Section 2 describes the method proposed in this paper, we propose a multi-view geometric enhancement strategy, introduce an efficient multi-scale feature fusion method, and design an attention-masking mechanism to improve model performance. Section 3 conducted experiments on different datasets and evaluated the methods proposed in this article. Finally, the conclusive remarks are given in Section 4.

2. Methods

The fittings detection method based on multi-scale geometric transformation and attention-masking mechanism (MGA-DETR) proposed in this paper is shown in Figure 4. The method is mainly divided into four parts: backbone, encoder, decoder, and prediction head. The backbone is used to extract image features and convert them into one-dimensional image sequences. In the encoder, the self-attention mechanism is used to obtain the relationship between image sequences, and then the trained image sequence features are output. The decoder initializes the object queries vector and is trained by the self-attention mechanism to learn the relationship between the object queries vector and image features. In the prediction header, a binary matching method is used to classify the category of the object queries vector and locate the position of the boundary box, completing the detection of transmission line fittings.

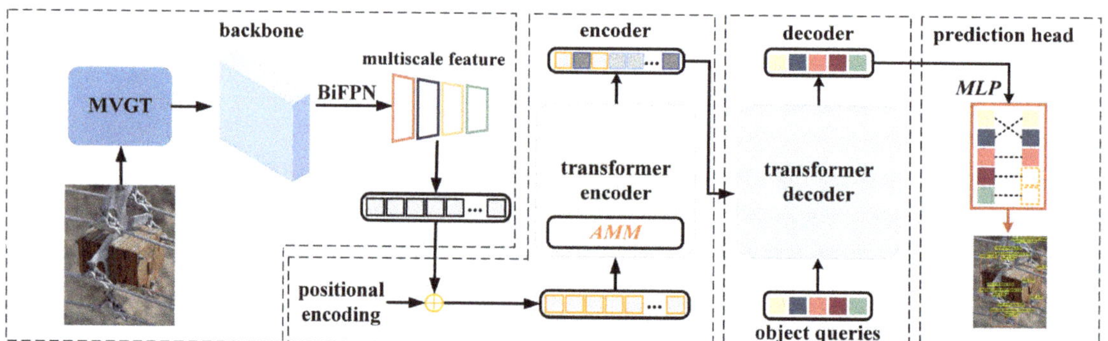

Figure 4. The basic architecture of the MAG-DETR.

Firstly, we designed a multi-view geometric transformation strategy (MVGT) to improve the detection performance of the model for fittings under different visual conditions in the backbone network part. Then, we introduced an efficient multi-scale feature fusion method (BiFPN) to improve the detection accuracy of the model for objects with different scales. Finally, to reduce the computational complexity of the model and achieve efficient transmission line inspection, this paper introduces an attention-masking mechanism (AMM). This method improves model detection by designing a scoring mechanism to filter out image regions that are less relevant to model detection.

2.1. Multi-View Geometric Transformation Strategy

When the distribution of test samples and training samples is different, the performance of object detection will decrease. There are many reasons for this problem, such as changes in the surface of objects under different lighting or weather conditions. Most methods to solve this problem focus on obtaining more data to enrich the feature representation of the object. In the field of object detection, there are usually two ways to obtain richer image feature representations. One method uses models to generate virtual images and add them to the dataset to increase the amount of data [31–33]. The other method uses methods such as random clipping and horizontal flipping to obtain high-quality feature representations during data preprocessing [34–36]. However, these methods do not pay attention to the geometric changes of the object caused by different shooting angles. This problem is particularly prominent in the inspection of power transmission lines. When the drone is shooting from different angles of view, the appearance of fittings can signifi-cantly change, leading to missed inspections and false inspections. Based on the above reasons, as shown in Figure 5, we propose the MVGT module that uses homomorphic transformation to bridge the gap between objects caused by geometric changes, and then fuses image features to improve the detection performance of fittings at different shooting angles.

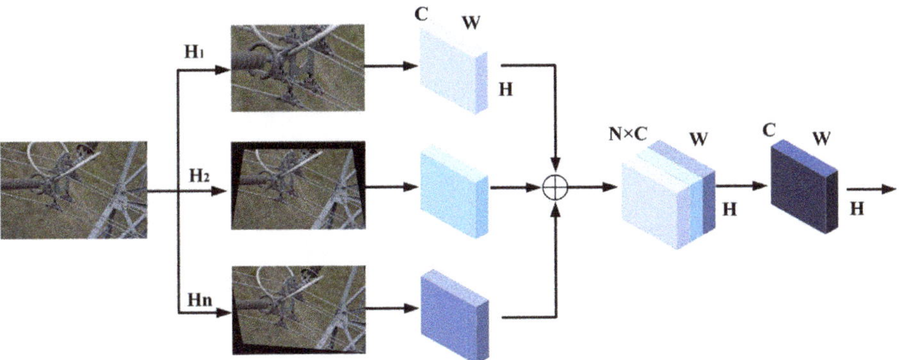

Figure 5. The architecture of the module of MVGT.

The homography transformation is a two-dimensional projection transformation that maps a point in one plane to another plane. Here, a plane refers to a planar surface in a two-dimensional image. The mapping relationship of corresponding points becomes the homography matrix. The calculation method is as follows:

$$(x_i, y_i, w_i)^T = H_i \times (x_{i'}, y_{i'}, w_{i'})^T \tag{1}$$

where x_i, y_i are the horizontal and vertical coordinates of the original image, and $x_{i'}, y_{i'}$ are the horizontal and axial coordinates of the image after the homography transformation. Set $w_i = w_{i'} = 1$ as the normalization point. H_i is a 3×3 homography matrix, it can be expressed as follows:

$$H_i = \begin{pmatrix} h_{00} & h_{01} & h_{02} \\ h_{10} & h_{11} & h_{12} \\ h_{20} & h_{21} & h_{22} \end{pmatrix} \tag{2}$$

So the $x_{i'}$ and $y_{i'}$ can be calculated by the following:

$$x_{i'} = \frac{h_{00}x + h_{01}y + h_{02}}{h_{20}x + h_{21}y + h_{22}} \tag{3}$$

$$y_{i'} = \frac{h_{10}x + h_{11}y + h_{12}}{h_{20}x + h_{21}y + h_{22}} \tag{4}$$

Therefore, when the coordinates of the four corresponding points are known, the homography matrix H_i can be obtained. In this paper, we have designed n sets of homography matrices to obtain corresponding homography-transformed images. After that, the homomorphic transformed image features are spliced to obtain features with the size of $H \times W \times NC$. Finally, we use a 1×1 convolution pair to reduce the dimension of the fused feature to the $H \times W \times C$ dimension. By combining the image features after homography transformation, the model can learn pixel changes from different perspectives, further improving the detection performance of fittings in transmission lines.

2.2. Bidirectional Feature Pyramid Network

UAVs fly high in the sky with a wide field of vision. The transmission line images captured by UAVs contain multiple categories of fittings. As shown in Figure 2, the range of fittings scales in different datasets are widely distributed. In the inspection of transmission lines, it is often due to the low resolution of small-size fittings, the missing details of the fittings, and the lack of features that can be extracted, which can easily lead to issues such as missing inspection. Therefore, the detection of such fittings has become the focus and difficulty of research.

In object detection methods, feature pyramid networks (FPN) are mainly used to improve the detection ability of models for objects of different scales [37]. As shown in Figure 6a, the main idea of the FPN is to fuse the context information of image features, enhancing the representation ability of shallow feature maps, and improving the detection ability of small-scale objects. Aiming at the defect of only focusing on one direction of information flow in FPN, Liu et al. [38] proposed the PAFPN to further fuse image features of different scales by adding a bottom-up approach, as shown in Figure 6b. In this paper, we introduce a bidirectional feature pyramid network (BiFPN) to optimize multiscale feature fusion in a more intuitive and principled manner [39], as shown in Figure 6c.

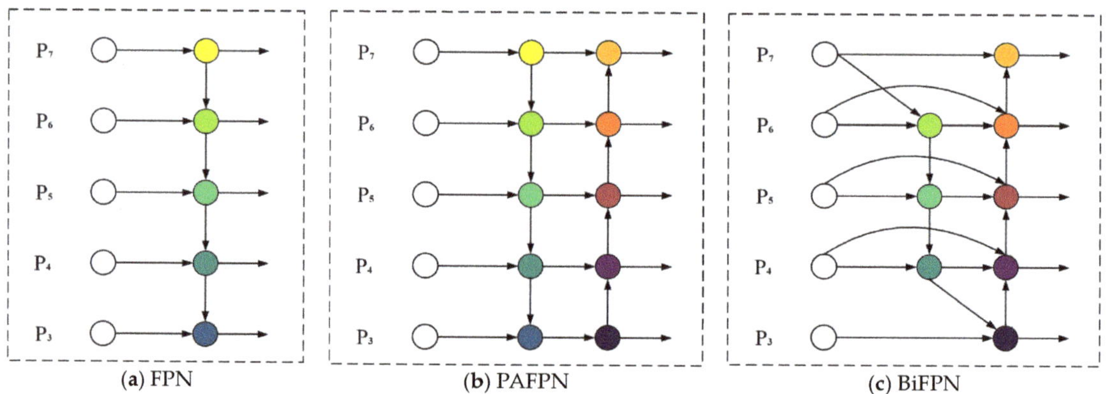

(a) FPN   (b) PAFPN   (c) BiFPN

Figure 6. The architectures of different FPNs.

First, assume that there is a set of image features $P_i \in \{P_1, P_2, \ldots, P_n\}$ with different scales. Where P_i represents the image features of the i level resolution. Effective multiscale feature extraction can be considered as a process in which P_i fuses different resolution features through a special spatial transformation function, with the ultimate goal of achieving feature enhancement. The fusion process is shown in Figure 6a, in which the network uses image features at levels 3 to 7, with the feature resolution at the level i being $1/2^i$ times the input image resolution.

BiFPN adopts a bidirectional feature fusion idea that combines top-down and bottom-up. In the top-down process, the seventh level node is deleted, which only has a single resolution input and has a small contribution to feature multiscale fusion. Deleting this node can simplify the network structure. At the same time, combining a top-down route with a bottom-up route increases the hierarchical resolution information required for the scale fusion process with minimal operational costs. Unlike the FPN, which only performs one feature fusion operation, the BiFPN regards the fusion process as an independent network module, connecting multiple feature fusion modules in series to achieve more possible fusion results.

In the top-down and bottom-up routes, upper and lower sampling methods are used to adjust the size of the feature map to be consistent, and a fast normalized feature fusion algorithm is used to fuse the adjusted feature map. The basic idea of a fast normalized feature fusion algorithm is that each target to be identified has its specificity, such as diverse scales and complex backgrounds. Therefore, visual features of different scales have different contributions to the network detection of the object. This paper uses learnable scalar values to measure the contribution of different levels of resolution features to the final prediction of the network. Using the softmax function to limit scalar values is a good method, but softmax can significantly reduce the GPU processing speed. To achieve acceleration, using a direct normalization algorithm can solve this problem:

$$w_i' = \frac{w_i}{\varepsilon + \sum_j w_j} \tag{5}$$

where ε is a minimum value. In order to avoid numerical instability that may occur during normalization calculations, we set $\varepsilon = 0.0001$. The w_i is the learned scalar value. To ensure $w_i \geq 0$, we use the ReLU activation function for each generated w_i.

The improved network uses three different scale features P_3, P_4 and P_5 extracted from the backbone network as inputs for cross-scale connectivity and weighted feature fusion. Take node P_5 as an example:

$$P_5^{t-d} = Conv\left(\frac{w_1 \cdot P_5 + w_2 \cdot Resize(P_6)}{w_1 + w_2 + \varepsilon}\right) \tag{6}$$

$$P_5^{b-u} = Conv\left(\frac{w_1' \cdot P_5 + w_2' \cdot P_5^{t-d} + w_3'Resize(P_4)}{w_1' + w_2' + w_3' + \varepsilon}\right) \tag{7}$$

where P_5^{t-d} is a top-down intermediate feature and P_5^{b-u} is a bottom-up output feature. Resize is an up-sampling operation or a down-sampling operation. Conv is a convolution operation.

2.3. Attention-Masking Mechanism

Although the model can obtain multiscale features of images using the BiFPN, there are still some problems. On the one hand, the self-attention mechanism in DETR can only process one-dimensional sequence data, and images belong to two-dimensional data. Therefore, when processing images, it is necessary to first perform dimensionality-reduction processing on the images. On the other hand, the image for object detection generally has a high resolution and mostly contains multiple targets at the same time. If the image is dimensionally reduced directly, the computational complexity of the Transformer codec will significantly increase. In order to solve this problem, in DETR, CNN is first used to extract image features and simultaneously reduce image dimensions, to control the overall calculation amount within an acceptable range. However, after using the BiFPN, the calculation amount of the model will be multiplied. To solve this problem, this paper introduces an attention-masking mechanism [40]. Firstly, a scoring network is used to predict the importance of the image sequence data input to the encoder, and the image sequence is trimmed hierarchically. Then, an attention-masking mechanism is used to prevent attention computation between the trimmed sequence data and other sequence data, thereby improving the computational speed.

The attention-masking mechanism designed in this paper is hierarchical, and as the calculation progresses, image sequence data with lower scores are gradually discarded. Specifically, we set a binary decision mask $S \in \{0,1\}^N$ to determine whether to discard or retain relevant data, where N is the length of the image sequence. When $S = 0$, it means that the data need to be discarded, but it is reserved anyway. During training, we initialize all S to 1 and gradually update S as the training progresses. Then, for the image sequence x in the input encoder, it is first passed into the MLP layer to obtain local features:

$$f^{local} = MLP(x) \tag{8}$$

Then, we interact with S on the local features of the image sequence to obtain the global features of the current image:

$$f^{global} = Agg(MLP(x), S) \tag{9}$$

where A can be obtained by simple averaging pooling:

$$Agg(f^{local}, S) = \frac{\sum\limits_{i=1}^{N} S_i f_i^{local}}{\sum\limits_{i=1}^{N} S_i} \tag{10}$$

Local features contain information about specific data in an image sequence, while global features contain all contextual information about the image. Therefore, we combine the two and transmit them to another MLP layer to obtain the probability of discarding or retaining image sequence data:

$$s = Softmax\left[MLP(f^{local}, f^{global})\right] \tag{11}$$

Subsequently, in order to maintain the length of the input image sequence during the training process unchanged while canceling the attention interaction between the trimmed sequence data and the data therein, we designed an attention-masking mechanism (AMM). To put it simply, AMM is added to attention calculation:

$$e_{ij} = \frac{(x_i w_Q)(x_j w_K)^T}{\sqrt{d}} \tag{12}$$

$$G_{ij} = \begin{cases} 1 & i = j \\ 0 & i \neq j \end{cases} \tag{13}$$

$$a_{ij} = \frac{\exp(e_{ij})G_{ij}}{\sum\limits_{k=1}^{n} \exp(e_{ik})G_{ij}} \tag{14}$$

where x is the data in the image sequence, w_Q and w_k are learnable parameter matrix, and d is used for normalization processing.

3. Experimental Results and Analysis

We trained the model using AdamW [41], setting the learning rate of the initial Transformer to 0.0001, the learning rate of the backbone network to 0.00001, the weight attenuation to 0.001, and the batch size to 8. The training process adopts the cosine annealing algorithm. When the detection accuracy of the validation set no longer increases, the learning rate is reduced by 10% until the learning rate accuracy no longer increases through adjustment. For the hyperparameter in the experiment, we set the number of object queries vectors to 100, and the number of layers of the Transformer encoder–decoder to 6. The experimental part was implemented using the Python framework and trained and tested using an NVIDIA Geforce GTX Titan device with four GPUs.

3.1. The Introduction of Datasets

In recent years, aerial photography technology has grown rapidly. To collect images of transmission lines, an aerial unmanned aerial vehicle (UAV) is not only simple to operate, but also can collect information quickly and safely. We used the UAVs aerial photography technology to obtain a large number of images of power transmission lines. The UAVs are equipped with a high-definition image transmission system, which can capture high-definition images of power transmission lines. Due to the different depth of field in the imaging of transmission line images captured by UAVs, we constructed three datasets in the experiment to verify the performance of the model.

(1) Fittings Datasets-25 (FD-25): Based on the progress of current UAV shooting technology, we constructed the fittings dataset of high-definition transmission line images captured by UAVs at ultra-wide angles. The characteristic of this dataset is that it has a

wide shooting range and contains a large number of fittings. We annotated the images according to the MS-COCO 2017 dataset's annotation specifications. The dataset includes a total of 4380 images and 50,830 annotation boxes. It includes 25 annotation categories, namely triangle yoke plate, right angle hanging board, u-type hanging ring, adjusting board, hanging board, towing board, sub-conductor spacer, shielded ring, grading ring, shock hammer, pre-twisted suspension class, bird nest, glass insulator without coating, compression tension class, suspension class, composite insulator, bowl hanging board, ball hanging ring, yoke plate, weight, extension rod, glass insulator with coating, lc-type yoke plate, upper-level suspension clamp, and interphase spacer. To our knowledge, the Fittings Dataset-25 currently contains the largest number of fittings components in the power industry and has the most detailed classification of fittings categories. An example image of the dataset is shown in Figure 7a,e.

Figure 7. Images from different datasets.

(2) Fittings Datasets-12 (FD-12): In addition to the transmission line images captured by UAVs at ultra-wide angles, we also annotated the relatively close-range transmission line images captured by UAVs. The datasets included 1,586 images and 10,185 annotation boxes. This includes 12 categories of fittings, including pre-twisted suspension clamp, bag-type suspension clamp, shielded ring, grading ring, spacer, wedge-type strain clamp, shockproof hammer, hanging board, weight, parallel groove clamp, u-type hanging ring, and yoke plate. Compared to the Fittings Datasets-25, the Fittings Datasets-12 has shorter shooting distances, fewer types of fittings, and a relatively rough classification of fittings. The image of the datasets is shown in Figure 7b,f.

(3) Fittings Datasets-9 (FD-9): There are a considerable number of small-scale fittings in transmission lines. Taking bolts as an example, the proportion of bolts in transmission line images is very small; usually, only a few pixel sizes; which leads to low accuracy of bolt recognition in object detection models. In response to the above issues, this paper cropped the Fittings Datasets-25 and Fittings Datasets-12, saving the areas with more small-scale fittings as new images and annotating them to increase the proportion of small-scale fittings in the input images. The dataset includes 1,800 images and 18,034 annotation boxes. This includes nine types of fittings: bolt, pre-twisted suspension clamp, u-type hanging ring, hanging board, adjusting board, bowl head hanging board, bag-type suspension clamp, yoke plate, and weight. An example image of the dataset is shown in Figure 7c,g.

3.2. Comparative Experiment

To verify the effectiveness of the proposed method in the fittings detection of transmission lines, we first conducted experiments using different models in the datasets constructed in this paper. As shown in Table 1, the AP is the average accuracy of the model detecting all labels in the datasets. GFLOPs are Giga Floating point Operations Per Second, FPS is the number of frames transmitted per second, and params is the number of parameters for the model.

Table 1. Experimental results of different fittings datasets.

Model	AP (FD-9)	AP (FD-12)	AP (FD-25)	GFLOPs/FPS	Params
Faster R-CNN	80.2	75.1	59.4	246/20	60 M
YOLOX	83.4	78.3	61.3	**73.8/81.3**	**25.3 M**
DETR	85.6	78.6	61.7	86/28	41 M
Deformable DETR	85.9	81.2	62.5	173/19	40 M
Sparse DETR	86.2	81.5	63.2	113/21.2	41 M
MGA-DETR	**88.7**	**83.4**	**66.8**	101/25.7	38 M

From Table 1, it can be seen that in the three types of fittings datasets, the MGA-DETR proposed in this paper achieves the highest average precision (AP) in fittings-detecting transmission lines. In the fittings datasets-9, the AP of MGA-DETR reached 88.7%, an increase of 3.1% compared to the baseline model DETR. In the fittings datasets-12, the AP value of MGA-DETR reached 83.4%, an increase of 4.8% compared to the baseline model DETR. In fittings datasets-25, the AP value of MGA-DETR reached 66.8%, an increase of 5.1% compared to the baseline model DETR. Compared to the three types of datasets, the detection accuracy of the fittings datasets-25 is relatively low because the images in the dataset are taken at ultra-wide angles, and the same image contains a variety of fittings types with significant scale changes. Through experiments, it has been proven that the model proposed in this article is of great help for the fittings detection of transmission lines. Comparing the params of different models, it can be found that the YOLOX has the smallest params. YOLOX is a single-stage object detection model. YOLOX introduces anchor-free, greatly reducing computational complexity while avoiding anchor-parameter tuning. Therefore, it has relatively large advantages in GFLOPs, FPS, and params. The method proposed in this paper is based on the transformer, and due to the self-attention mechanism in the transformer, the computational complexity of the model is relatively large. Compared to other methods based on transformer, our method introduces AMM, which successfully accelerates the calculation speed of the model and reduces the number of parameters in it. The MGA-DETR proposed in this paper has improved the params and FPS of the Deformable DETR, which also uses FPN, further proving the effectiveness of the proposed method.

Figure 8 shows the detection performance of the proposed method in different fittings datasets. Among them, Figure 8a,d show the detection performance of Fittings Datasets-25, Figure 8b,e show the detection performance of Fit tings Datasets-12, and Figure 8c,f show the detection performance of Fittings Datasets-9. From the figure, it can be seen that the method proposed in this article effectively detects the presence of fittings in the image in all three types of datasets. Taking Figure 8b,e as examples, the shape of the bag-type suspension clamp in the image has undergone significant changes due to different shooting angles. However, the method in this paper accurately detects two different shapes of bag-type suspension clamps. This further proves the effectiveness of the MAGT module proposed in this paper.

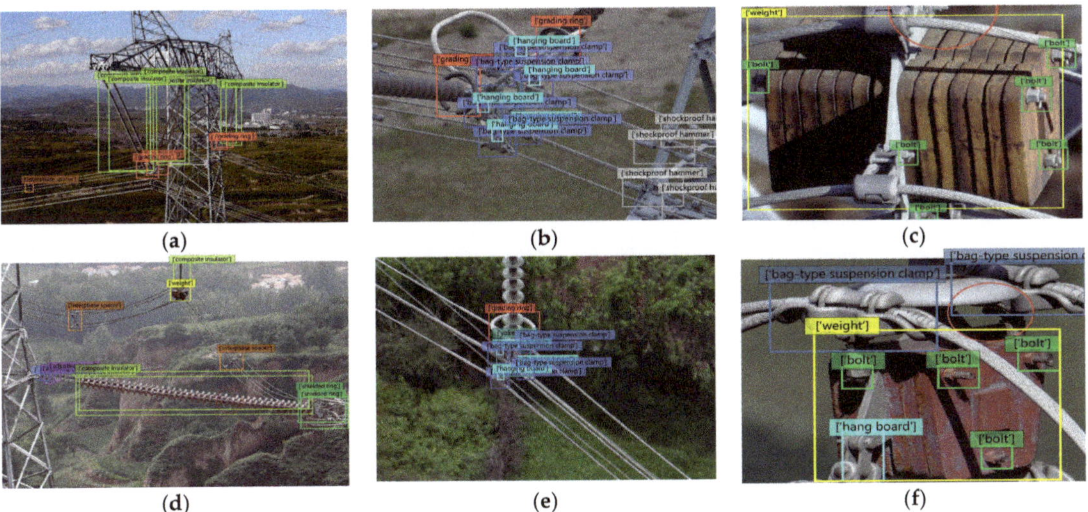

Figure 8. The architectures of different FPN.

Table 2 shows the detection results of fittings at different scales in three datasets. Among them, the glass insulator, grading ring, and shielded ring are large-scale fittings; the adjusting board, yoke plate, and weight are mesoscale fittings; and the hanging board, bowl hanging board, and u-type hanging ring are small-scale fittings. The × symbols in Table 2 indicate that the dataset does not contain fittings of this category. Through comparison, it can be seen that our proposed MGA-DETR has better performance in fittings detection at different scales. Taking the small-scale fittings hanging board as an example, the AP in three datasets was 86.9%, 80.4%, and 63.1%, respectively. Compared with the baseline model, the DETR increased by 7.2%, 4.5%, and 9.7%, respectively. The experiment shows that the introduction of the BiFPN in DETR has better detection performance for different scales of fittings.

Table 2. Experimental results of DETR/MGA-DETR on different categories in three datasets.

Fittings	AP (FD-9)	AP (FD-12)	AP (FD-25)
glass insulator	×	×	×
grading ring	×	83.1/**89.7**	72.6/**80.4**
shielded ring	×	83.2/**90.2**	69.8/**79.5**
adjusting board	87.3/**90.7**	78.8/**85.1**	57.9/**68.7**
yoke plate	87.9/**91.2**	79.3/**84.4**	58.3/**69.1**
weight	88.2/**91.3**	78.2/**85.2**	57.5/**68.2**
hanging board	79.7/**86.9**	75.9/**80.4**	53.4/**63.1**
bowl hanging board	81.3/**86.6**	76.1/**80.5**	52.7/**62.9**
u-type hanging ring	82.6/**86.9**	75.4/**80.1**	53.5/**62.3**

3.3. Ablation Experiment

In this section, we designed a series of ablation experiments to demonstrate the effectiveness of each module used in this paper. We used the Fittings Datasets-12 with moderate shooting distance and relatively rich fittings categories to verify the AP of the model.

As shown in Table 3, we analyzed the impact of different module combinations on the experimental results. When all three models are not used, the AP at this time is 78.6%. When only the MVGT module is used, the AP of the model increases by 1.5%, indicating that the feature combination after image homography transformation is beneficial for detecting fittings under different visual conditions. When only the BiFPN is used, the AP of the model increases by 1.8%, indicating that multi-scale feature fusion is more effective in transmission line images with significant scale changes. When only using the AMM module, the AP of the model increases by 1.1%, indicating that the model can improve detection accuracy by filtering out irrelevant background information. When three modules are added simultaneously, the AP reaches its maximum.

Table 3. The impact of different modules on experimental results.

Model	MVGT	BiFPN	AMM	AP (FD-9)	AP (FD-12)	AP (FD-25)
	×	×	×	85.6	78.6	61.7
	√	×	×	85.9	80.1	63.2
	×	√	×	86.3	80.4	63.9
MGA-DETR	×	×	√	85.8	79.7	62.9
	√	√	×	87.6	82.9	65.4
	√	×	√	87.3	81.6	64.7
	×	√	√	87.4	81.7	64.9
	√	√	√	**88.7**	**83.4**	**66.8**

In Table 4, we analyzed in detail the impact of different numbers of homography transformations on model performance. When the number is 0, the AP of the model is only 81.7%. With the fusion of image features after homography transformation, the model performance reaches its optimal level at the number of 4, with an AP of 83.4%. When the number of homomorphic transformations further increases, the model performance decreases, indicating that the model has fully learned the geometric transformations in different views at this time. Our analysis concludes that the reason is that with the increase in the number, the model overfitting will lead to a decrease in AP.

Table 4. The influence of different numbers of homography transformations on experimental results.

Model	Number	AP (FD-9)	AP (FD-12)	AP (FD-25)
	0	87.4	81.7	64.9
	1	87.8	82.5	65.3
	2	88.0	82.9	65.9
MVGT	3	88.3	83.1	66.5
	4	**88.7**	**83.4**	**66.8**
	5	88.6	83.3	66.6
	6	88.1	82.7	66.1

As shown in Table 5, we analyzed the impact of different FPNs on model performance. When FPN is not used, the model's AP is only 81.6%. When using FPN, the AP increased by 0.6%, indicating that learning multi-scale image features helps the model detect transmission line fittings at different scales. However, FPN only considers the top-down feature fusion, while PAFPN considers the bottom-up feature fusion on this basis. However, the efficiency of the two feature-fusion methods did not reach the optimal level. In this article, we introduced the BiFPN, which further improved the AP of the model, demonstrating the effectiveness of our method.

Table 5. The Influence of Different FPNs on Experimental Results.

Model	FPN	PAFPN	BiFPN	AP (FD-9)	AP (FD-12)	AP (FD-25)
MGA-DETR	×	×	×	85.1	81.6	60.7
	√	×	×	86.7	82.3	62.1
	×	√	×	87.2	82.8	64.3
	×	×	√	**88.7**	**83.4**	**66.8**

4. Conclusions

In order to improve the accuracy of transmission line fittings detection, this paper proposes a fittings detection method based on multi-scale geometric transformation and attention-masking mechanism. Firstly, we designed an MVGT module to utilize homography transformation to obtain image features from different views. Then, the BiFPN was introduced to efficiently fuse multi-scale features of images. Finally, we used an AMM module to improve model speed by masking the attention interaction between image sequence data with lower scores and other data. This paper constructs three different datasets of transmission line fittings and conducts experiments on them. The experimental results show that the proposed method effectively improves the performance of transmission line fittings detection. In the next step of our work, we will study the deployment of the model to obtain its application in the industry.

Author Contributions: Conceptualization, N.W. and J.Z.; methodology, N.W. and K.Z.; software, L.Z.; validation, Z.H., X.W. and Y.Z.; data curation, N.W., K.Z. and W.L.; writing—original draft preparation, N.W.; writing—review and editing, J.Z. and W.L. All authors have read and agreed to the published version of the manuscript.

Funding: This research received no external funding.

Institutional Review Board Statement: Not applicable.

Informed Consent Statement: Not applicable.

Data Availability Statement: Not applicable.

Conflicts of Interest: The authors declare no conflict of interest. The funders had no role in the design of the study; in the collection, analysis, or interpretation of data; in the writing of the manuscript; or in the decision to publish the results.

Abbreviations

The following abbreviations are used in this manuscript:

UAV	Unmanned Aerial Vehicle
CNN	Convolutional Neural Network
MVGT	Multi-View Geometric Transformation strategy
BiFPN	Bidirectional Feature Pyramid Network
AMM	Attention-Masking Mechanism

References

1. Dong, Z.; Zhao, H.; Wen, F.; Xue, Y. From Smart Grid to Energy Internet: Basic Concept and Research Framework. *Autom. Electr. Power Syst.* **2014**, *15*, 1–11.
2. Nguyen, V.; Jenssen, R.; Roverso, D. Automatic autonomous vision-based power line inspection: A review of current status and the potential role of deep learning. *Int. J. Electr. Power Energy Syst.* **2018**, *99*, 107–120. [CrossRef]
3. Zhao, Z.; Zhang, W.; Zhai, Y.; Zhao, W.; Zhang, K. Concept, Research Status and Prospect of Electric Power Vision Technology. *Electr. Power Sci. Eng.* **2020**, *57*, 57–69.
4. Cheng, Z.; Fan, M.; Li, Y.; Zhao, Y.; Li, C. Review on Semantic Segmentation of UAV Aerial Images. *Comput. Eng. Appl.* **2021**, *57*, 57–69.

5. Deng, C.; Wang, S.; Huang, Z. Unmanned aerial vehicles for power line inspection: A cooperative way in platforms and communications. *J. Commun.* **2014**, *9*, 687–692. [CrossRef]

6. Hu, B.; Wang, J. Deep learning based on hand gesture recognition and UAV flight controls. *Int. J. Autom. Comput.* **2020**, *17*, 17–29. [CrossRef]

7. Zhao, Z.; Cui, Y. Research progress of visual detection methods for transmission line key components based on deep learning. *Electr. Power Sci. Eng.* **2018**, *34*, 1.

8. Girshick, R.; Donahue, J.; Darrell, T. Rich feature hierarchies for accurate object detection and semantic segmentation. In Proceedings of the IEEE Conference on Computer Vision and Pattern Recognition, Columbus, OH, USA, 23–28 June 2014; pp. 580–587.

9. Girshick, R. Fast r-cnn. In Proceedings of the IEEE International Conference on Computer Vision, Santiago, Chile, 7–13 December 2015; pp. 1440–1448.

10. Ren, S.; He, K.; Girshick, R. Faster r-cnn: Towards real-time object detection with region proposal networks. *IEEE Trans. Pattern Anal. Mach. Intell.* **2015**, *39*, 1137–1149. [CrossRef]

11. Sun, P.; Zhang, R.; Jiang, Y. Sparse r-cnn: End-to-end object detection with learnable proposals. In Proceedings of the Conference on Computer Vision and Pattern Recognition, Online, 19–25 June 2021; pp. 14454–14463.

12. Liu, W.; Anguelov, D.; Erhan, D. SSD: Single shot multibox detector. In Proceedings of the European Conference on Computer Vision, Amsterdam, The Netherlands, 10–16 October 2016; pp. 21–37.

13. Redmon, J.; Divvala, S.; Girshick, R. You only look once: Unified, real-time object detection. In Proceedings of the IEEE Conference on Computer Vision and Pattern Recognition, Las Vegas, NV, USA, 26 June–1 July 2016; pp. 779–788.

14. Ge, Z.; Liu, S.; Wang, F. Yolox: Exceeding Yolo Series in 2021. *arXiv* **2021**, arXiv:2107.08430.

15. Salman, K.; Muzammal, N.; Munawar, H. Transformers in Vision: A Survey. *ACM Comput. Surv. (CSUR)* **2022**, *54*, 1–41.

16. Carion, N.; Massa, F.; Synnaeve, G. End-to-end object detection with transformers. In Proceedings of the European Conference on Computer Vision, Online, 23–28 August 2020; pp. 213–229.

17. Zhu, X.; Su, W.; Lu, L. Deformable DETR: Deformable Transformers for End-to-End Object Detection. *arXiv* **2021**, arXiv:2010.04159.

18. Roh, B.; Shin, J.; Shin, W. Sparse DETR: Efficient End-to-End Object Detection with Learnable Sparsity. *arXiv* **2021**, arXiv:2111.14330.

19. Fang, Y.; Liao, B.; Wang, X. You only look at one sequence: Rethinking transformer in vision through object detection. *Adv. Neural Inf. Process. Syst.* **2021**, *34*, 26183–26197.

20. Song, H.; Sun, D.; Chun, S. ViDT: An Efficient and Effective Fully Transformerbased Object Detector. *arXiv* **2021**, arXiv:2110.03921.

21. Wu, K.; Peng, H.; Chen, M. Rethinking and improving relative position encoding for vision transformer. In Proceedings of the International Conference on Computer Vision, Montreal, Canada, 10–17 October 2021; pp. 10033–10041.

22. Qi, Y.; Wu, X.; Zhao, Z.; Shi, B.; Nie, L. Bolt defect detection for aerial transmission lines using Faster R-CNN with an embedded dual attention mechanism. *J. Image Graph.* **2021**, *26*, 2594–2604.

23. Zhang, S.; Wang, H.; Dong, X. Bolt Detection Technology of Transmission Lines Based on Deep Learning. *Power Syst. Technol.* **2020**, *45*, 2821–2829.

24. Zhong, J.; Liu, Z.; Han, Z. A CNN-based defect inspection method for catenary split pins in high-speed railway. *IEEE Trans. Instrum. Meas.* **2019**, *68*, 2849–2860. [CrossRef]

25. Zhao, Z.; Duan, J.; Kong, Y.; Zhang, D. Construction and Application of Bolt and Nut Pair Knowledge Graph Based on GGNN. *Power Syst. Technol.* **2021**, *56*, 98–106.

26. Zhao, Z.; Xu, G.; Qi, Y. Multi-patch deep features for power line insulator status classification from aerial images. In Proceedings of the International Joint Conference on Neural Networks, Vancouver, BC, Canada, 24–29 July 2016; pp. 3187–3194.

27. Zhao, Z.; Ma, D.; Ding, J. Weakly Supervised Detection Method for Pin-missing Bolt of Transmission Line Based on SAW-PCL. *J. Beijing Univ. Aeronaut. Astronaut.* **2023**, 1–10. [CrossRef]

28. Zhang, K.; Zhao, K.; Guo, X. HRM-CenterNet: A High-Resolution Real-time Fittings Detection Method. In Proceedings of the International Conference on Systems, Man, and Cybernetics, Melbourne, Australia, 17–20 October 2021; pp. 564–569.

29. Zhang, K.; He, Y.; Zhao, K. Multi Label Classification of Bolt Attributes based on Deformable NTS-Net Network. *J. Image Graph.* **2021**, *26*, 2582–2593.

30. Lou, W.; Zhang, K.; Guo, X. PAformer: Visually Indistinguishable Bolt Defect Recognition Based on Bolt Position and Attributes. In Proceedings of the Asia-Pacific Signal and Information Processing Association Annual Summit and Conference, Chiang Mai, Thailand, 7–10 November 2022; pp. 884–889.

31. Qi, Y.; Lang, Y.; Zhao, Z.; Jiang, A.; Nie, L. Relativistic GAN for bolts image generation with attention mechanism. *Electr. Meas. Instrum.* **2019**, *56*, 64–69.

32. Yu, Y.; Gong, Z.; Zhong, P. Unsupervised representation learning with deep convolutional neural network for remote sensing images. In Proceedings of the Image and Graphics: 9th International Conference, Los Angeles, CA, USA, 28–30 July 2017; pp. 97–108.

33. Ledig, C.; Theis, L.; Huszar, F. Photo-Realistic Single Image Super-Resolution Using a Generative Adversarial Network. In Proceedings of the IEEE Conference Computer Vision and Pattern Recognition, Hawaii, HI, USA, 21–26 July 2017; pp. 4681–4690.

34. He, K.; Zhang, X.; Ren, S.; Sun, J. Deep residual learning for image recognition. In Proceedings of the IEEE Conference on Computer Vision and Pattern Recognition, Las Vegas, NV, USA, 26 June–1 July 2016; pp. 770–778.

35. He, J.; Chen, J.; Liu, S. TransFG: A Transformer Architecture for Fine-Grained Recognition. *arXiv* **2021**, arXiv:2103.07976. [CrossRef]
36. Chen, Z.; Wei, X.; Wang, P.; Guo, Y. Multi-Label Image Recognition with Graph Convolutional Networks. In Proceedings of the IEEE Conference on Computer Vision and Pattern Recognition, Long Beach, CA, USA, 16–20 June 2019; pp. 5177–5186.
37. Lin, T.; Dollar, P.; Girshick, R. Feature pyramid networks for object detection. In Proceedings of the IEEE Conference on Computer Vision and Pattern Recognition, Hawaii, HI, USA, 21–26 July 2017; pp. 2117–2125.
38. Liu, S.; Qi, L.; Qin, H. Path aggregation network for instance segmentation. In Proceedings of the IEEE Conference on Computer Vision and Pattern Recognition, Salt Lake City, UT, USA, 18–22 June 2018; pp. 8759–8768.
39. Tan, M.; Pang, R.; Le, Q. Efficientdet: Scalable and efficient object detection. In Proceedings of the IEEE Conference on Computer Vision and Pattern Recognition, Seattle, WA, USA, 14–19 June 2020; pp. 10781–10790.
40. Rao, Y.; Zhao, W.; Liu, B. DynamicViT: Efficient Vision Transformers with Dynamic Token Sparsification. *Adv. Neural Inf. Process. Syst.* **2021**, *34*, 13937–13949.
41. Loshchilov, I.; Hutter, F. Decoupled Weight Decay Regularization. *arXiv* **2018**, arXiv:1711.05101.

Article

A New Orientation Detection Method for Tilting Insulators Incorporating Angle Regression and Priori Constraints

Jianli Zhao *, Liangshuai Liu, Ze Chen, Yanpeng Ji and Haiyan Feng

Electric Power Research Institute, State Grid Hebei Electric Power Co., Ltd., Shijiazhuang 050017, China
* Correspondence: zhaojianli0825@163.com; Tel.: +86-18503288290

Abstract: The accurate detection of insulators is an important prerequisite for insulator fault diagnosis. To solve the problem of background interference and overlap caused by the axis-aligned bounding boxes in the tilting insulator detection tasks, we construct an improved detection architecture according to the scale and tilt features of the insulators from several perspectives, such as bounding box representation, loss function, and anchor box construction. A new orientation detection method for tilting insulators based on angle regression and priori constraints is put forward in this paper. Ablation tests and comparative validation tests were conducted on a self-built aerial insulator image dataset. The results show that the detection accuracy of our model was increased by 7.98% compared with that of the baseline, and the overall detection accuracy reached 82.33%. Moreover, the detection effect of our method was better than that of the YOLOv5 detection model and other orientation detection models. Our model provides a new idea for the accurate orientation detection of insulators.

Keywords: tilting insulator; orientation detection; angle regression; prior constraint

1. Introduction

As an essential component of transmission lines, insulators undertake the functions of electrical insulation and structural support [1]. Under multiple impacts of high voltage, mechanical stress, and harsh environment, insulators are prone to defects such as fouling, flashing, breakage, and string dropping. In this case, the fast and accurate detection of insulators and their defects has become an essential task to ensure the safety of transmission lines [2,3]. Many relevant studies focused on the accuracy and speed of detection [4,5], but an equally important issue, how to accurately recognize those tilting insulators, still needs exploring.

Insulator detection methods [6–11] can be roughly classified into two categories: digital-image-processing-based methods and deep-learning-based methods. The insulator contamination detection method proposed by Xun et al. [6] is a typical method using digital-image processing technology. It improves the watershed algorithm by similar-region fusion and minimization segmentation and effectively avoids the over-segmentation phenomenon. Zhai et al. [7] introduced airspace morphological consistency features to obtain high-accuracy insulator pinpointing in the insulator detection task. Zhang et al. [8] generated feature sequences by various texture extraction methods and achieved good insulator-defect detection results. In the absence of background interference, insulator detection methods based on digital-image processing have high detection accuracy, but they rely too much on artificially designed features and have poor robustness, which makes them struggle to handle aerial images taken by unmanned aerial vehicles (UAV) with complex background environments and small insulator targets. With the development of UAV image-acquisition technology and object detection technology in electric power field, deep learning based insulator and defect detection methods have been widely investigated. For example, Wang et al. [9] combined a two-stage insulation anomaly detection model with a few-sample learning method to achieve high-precision defect detection. Following

Citation: Zhao, J.; Liu, L.; Chen, Z.; Ji, Y.; Feng, H. A New Orientation Detection Method for Tilting Insulators Incorporating Angle Regression and Priori Constraints. *Sensors* **2022**, *22*, 9773. https://doi.org/10.3390/s22249773

Academic Editors: Ke Zhang and Yincheng Qi

Received: 18 November 2022
Accepted: 2 December 2022
Published: 13 December 2022

the two-stage detection idea of combining the target detection task with the semantic segmentation task, Ling et al. [10] proposed a lightweight and high-precision insulator detection method. Li et al. [11] applied the YOLOv5 series model to the insulator detection task and achieved a detection accuracy of up to 96% and a detection speed of 42 images per second on the self-built insulator detection dataset, proving the superiority of the YOLOv5 series model.

Currently, great breakthroughs have been made in the study on the insulator detection of aerial insulator images. However, most of the methods are the simple migration of generic object detection to insulator detection. In some complex conditions such as dense mutual occlusion and dense distribution, which often result in unnecessary background noise and overlap, the insulator detection effect is not ideal, and sometimes there even exists a phenomenon of missing detection [12]. In aerial images shot by UAVs, insulators appear with different aspect ratios and tilt angles, while the general object detection models cannot fully utilize the scale and tilt features. When detecting insulator overlap, if the axis-aligned bounding boxes are too close to each other, the non-maximum suppression algorithm often fails, resulting in missing detection.

In this paper, we apply the YOLOv5 model to the insulator detection field and propose a tilting-insulator detection model based on angle regression and prior constraint of scale (RAPC-YOLO). In particular, we firstly introduce the angle regression loss in the loss function and combine the oriented bounding box with the YOLOv5 object detection model, thus effectively improving the tilting-insulator detection effect. Then, the anchor box parameters of the detection model are analyzed and optimized by the clustering algorithm according to the scale features of the insulators. Last but not least, we introduce a rotational uncertainty function to guide the learning of the angle regression loss according to the angle distribution of the insulator's oriented bounding box, so as to improve the robustness of the model to the tilting angle.

2. Materials and Methods

In this section, an overview of our proposed method is given. The architecture of the tilting insulator detection model is shown in Figure 1. The detailed optimization adjustments on the YOLOv5 network is given in Section 2.1. After that, a specific method to implement the scale priori constraint is presented in Section 2.2. Finally, the influence of the tilting angle is analyzed, and a method to obtain the angle priori constraints is put forward in Section 2.3.

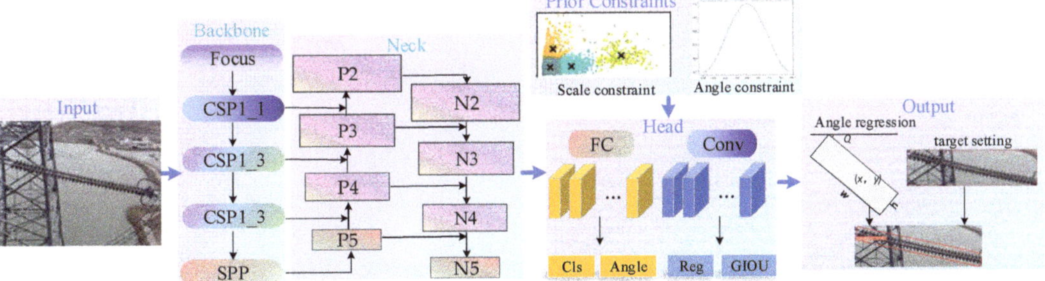

Figure 1. Architecture of RAPC-YOLO.

2.1. YOLOv5-Orientation Model

YOLOv5 is the fifth generation of the You Only Look Once (YOLO) series of single-stage detection models, and it has become one of the most popular baseline models in the field of target detection. To obtain the reliable detection of tilting insulators, we introduce the angle parameter to the axis-aligned bounding box to form an oriented bounding box adapted to the tilt characteristics of insulators. Then an angle regression loss is introduced into the loss function, so as to obtain a new YOLOv5-Orientation model adapted to the tilting-insulator detection task using loss descent learning.

The oriented bounding box parameter of the YOLOv5-Orientation model is defined as $[x, y, w, h, \theta]$, where (x, y) is the normalized centroid coordinates, w is the normalized short edge length, h is the normalized long edge length, and θ is the normalized tilt angle. The normalized tilt angle θ is derived from Equation (1).

$$\theta = \frac{Q}{90} \tag{1}$$

where Q is the tilt angle.

As shown in Figure 2, Q is the minimum angle required for the long side w of the rectangular box to coincide with the x-axis. If the rotation is clockwise, the tilt angle is positive; otherwise the tilt angle is negative. Thus, the value range of Q is $[-90°, 90°]$.

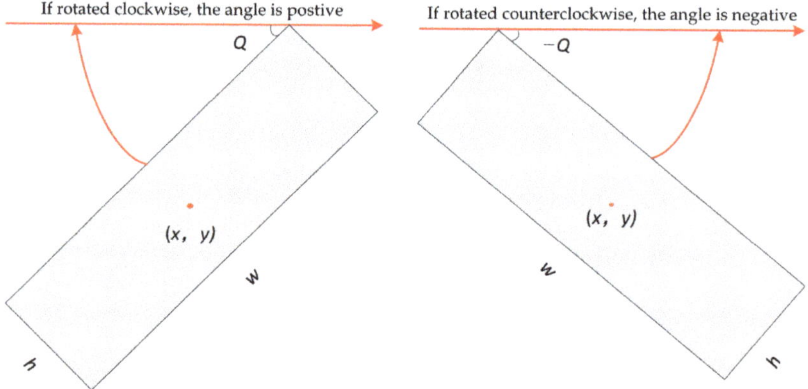

Figure 2. Schematic diagram of the oriented bounding box.

An angle regression loss based on the localization loss of the YOLOv5 model is introduced, consisting of the generalized intersection over union (GIOU) [13] loss and the smoothing loss [14], as expressed by Equation (2).

$$L_{reg}(o, l, g) = \sum_i \sum_{m \in S} o_{ij} SL_{GIOU}(l_i^m - g_j^{\Delta m}) + \sum_i o_{ij} \theta L_1(l_i^\theta - g_j^{\Delta \theta}) \tag{2}$$

where o denotes the label, l denotes the predicted oriented bounding box, g denotes the real oriented bounding box, o_{ij} is a binary variable and denotes the degree of matching between the label of the i-th default box and the label of the j-th real box, S is the set of parameters $\{x,y,w,h\}$, $g_j^{\Delta m}$ is the offset of parameters $\{x,y,w,h\}$, $g_j^{\Delta \theta}$ is the introduced angle offset, and the offset calculation formula is shown in Equation (3).

$$\begin{cases} g_j^{\Delta x} = (g_j^x - d_i^x)/d_i^w \\ g_j^{\Delta w} = \lg(\frac{g_j^w}{d_i^w}) \\ g_j^{\Delta y} = (g_j^y - d_i^y)/d_i^h \\ g_j^{\Delta h} = \lg(\frac{g_j^h}{d_i^h}) \\ g_j^{\Delta \theta} = (g_j^\theta - d_i^\theta) \end{cases} \tag{3}$$

The center coordinate offset $\left(g_j^{\Delta x}, g_j^{\Delta y}\right)$ is the normalized value of the difference between the center coordinates of the real rectangular box g and the default rectangular box d. The long side offset $g_j^{\Delta w}$ and the short side offset $g_j^{\Delta h}$ are the logarithm of the corresponding side-length ratio between the real rectangular box g and the default rectangular box d. The above four offsets are regressed by GIOU loss, and the formula is shown in Equation (4). The angle offset $g_j^{\Delta \theta}$ is the difference between the tilt angle g_j^θ of the real rectangular box g and the tilt angle d_j^θ of the default rectangular box d, and it is regressed by L_1-smoothing loss. The formula is shown in Equation (5).

$$L_{GIOU} = 1 - GIOU \tag{4}$$

$$smooth_{L_1}(m) = \begin{cases} 0.5\,m^2 & if\ |m| < 1 \\ |m| - 0.5 & otherwise \end{cases} \tag{5}$$

2.2. Scale Priori Constraints

Motivated by the optimization need of the initial anchor-box aspect ratio and number, the scale priori constraints are put forward. The preset parameters of the anchor boxes of the YOLOv5-Orientation model are extracted from the public dataset, and they do not match the insulator scale features. Therefore, in this paper, the K-means [15] clustering algorithm is used to cluster and analyze the scale parameters of each labeled oriented bounding box in the insulator dataset. The width–height ratio, size, and number of the optimized anchor box are used as the new anchor-box preset parameters. The specific implementation process is shown in Algorithm 1.

Algorithm 1 Overall process of anchor-box scale clustering

Input:
    The scale parameters of labeling box in the dataset, the maximum number of iterations I
Processing:
    for $m = 1;\ m \leq I;\ m{+}{+}$ **do**

1. The set of parameter samples obtained from the dataset $T = \{t_i | t_i \in R^v, i = 1, 2, 3, \ldots M\}$, t_i is a single sample, M is the number of labeled boxes, $v = 2$ is the parameter dimension, which is the width and height parameter, respectively.
2. Randomly initialize K samples as clustering centers to form the set of clustering centers $C(I) = \{c_j | c_j \in R^v\}$, c_j is a single clustering center, I is the number of iterations, and its initial value is 1.
3. Calculate the distance between each sample t_i of the sample set T and each cluster center c_j in $C(I)$ according to the Euclidean distance formula $d(t_i, c_j) = \sqrt{(t_i - c_j)^2}$ and merge each sample t_i into the cluster center c_j with the smallest Euclidean distance by the size of the Euclidean distance, i.e., $T_j = \{t_i | t_i \subseteq c_j\}$.
4. For each cluster T_j take the sample t_i belonging to it and calculate the new cluster center $\tilde{c}_j$ according to Equation $\tilde{c}_j = \sum_{t_i \in T_j} t_i / n$, and form the set of cluster centers $C(I+1)$ from the new cluster centers $\tilde{c}_j$.

    if $C(I+1) = C(I)$:
        end for
Output:
The width, height, and number parameters obtained by clustering

---

As an important hyperparameter of the K-means algorithm, K directly affects the clustering quality. In order to obtain the clustering results with high intra-cluster similarity and low inter-cluster similarity, CH is selected to measure the effects of different K-values on clustering quality, so as to obtain the best anchor-box preset parameters, calculated by Equation (6).

$$CH(K) = \frac{traceB/(K-1)}{traceW/(N-K)} \tag{6}$$

where, *traceB* denotes the trace of the inter-cluster dispersion matrix, K denotes the number of clustering centers, *traceW* denotes the trace of the intra-cluster dispersion matrix, N denotes the total number of records, and CH is proportional to the clustering quality.

2.3. Angle Priori Constraint

Studies [16–18] have shown that balanced sample distribution has a significant impact on detection performance. Therefore, the analysis of the insulator-tilting-angle samples is necessary, and the optimization of the unbalanced sample distribution can improve the robustness of the model to the tilt angle. Figure 3 shows the frequency distribution and probability density distribution of the tilt angle of the annotated boxes in the insulator dataset. The light blue histogram is the angular frequency, and the red curve is the fitted probability density.

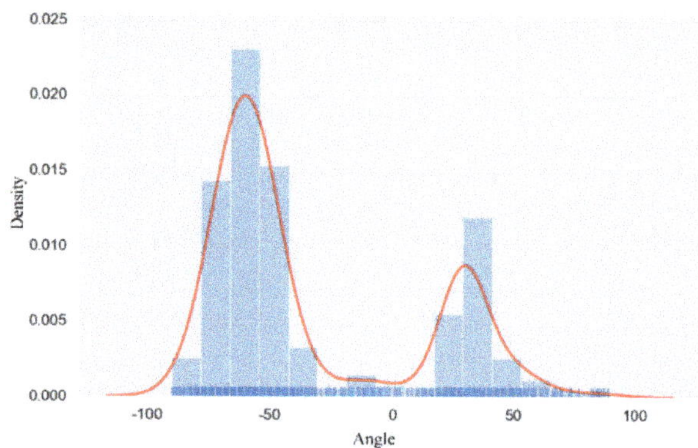

Figure 3. Probability density distribution of tilt angle in the insulator dataset.

It can be found that the overall distribution of the tilt angles is uneven. A large number of the tilt angles of these samples are concentrated around $-60°$ and $30°$. To reduce the effect of the uneven distribution of the tilt angle on the angle regression, we introduce a rotational uncertainty function [16] as a threshold function to control the regression loss, so as to obtain the angle prior constraint.

The formula of the rotational uncertainty function $D(\theta)$ is shown in Equation (7), where θ is the tilt angle, and δ is the angular hyperparameter when $D(\theta) = 0.5$. The visualization graph of the rotational uncertainty function is shown in Figure 4.

$$D(\theta) = \max(0.5, 1 + \frac{1 - \cos(4\theta)}{2\cos(4\delta) - 2}) \tag{7}$$

This function maps the tilt angle θ to the GIOU threshold and then controls the regression loss calculation by the GIOU threshold. In this way, the semantic features learned by the model in the interval with more distribution of tilt-angle samples can be migrated to the interval with less distribution of tilt-angle samples, so as to assist their detection. Herein, the GIOU threshold is set to 0.5 in reference to the threshold of anchor matching in the standard object-detection architecture.

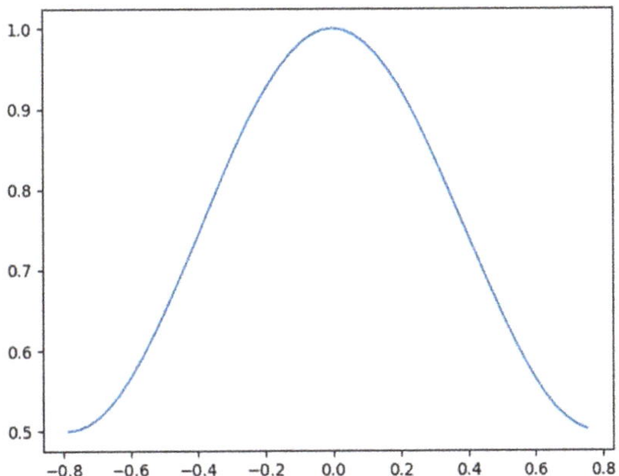

Figure 4. Visualization of rotational uncertainty function.

3. Test Results & Analysis

3.1. Test Data and Parameter Settings

The test dataset consists of insulator images taken by UAVs, including 1754 aerial images of insulators. Most images contain tilting insulators. About 1404 aerial-insulator images were randomly selected from the dataset to form a training set, while the remaining for a test set, and the ratio of the training set to the test set was 4:1.

In this paper, average precision (AP) is chosen as the test evaluation metric. It can reflect the comprehensive accuracy of each category and is derived by integrating the P–R curves constructed from recall and precision, as shown in Equation (8).

$$AP_m = \sum_m \int_0^1 P(r)d(r) \tag{8}$$

where $P(r)$ is the curve with recall as the independent variable and precision as the dependent variable; m is the GIOU threshold. Precision is the proportion of correctly predicted boxes, and recall is the proportion of predicted boxes among all of the real boxes. As shown in Equations (9) and (10), TP denotes the number of correctly detected targets, FP denotes the number of incorrectly detected targets, and FN denotes the number of unpredicted real boxes.

$$\text{Precision} = \frac{TP}{TP + FP} \tag{9}$$

$$\text{Recall} = \frac{TP}{TP + FN} \tag{10}$$

To verify the effectiveness of our proposed method for the tilting-insulator detection task, we adopt AP_{50} as the basic evaluation index and also select $AP_{50\sim95}$ as another evaluation index, which is more demanding for inspection.

The tests were conducted on an Ubuntu 18.04 operating system. The memory is 32GB; the graphics card is Nvidia GeForce RTX2080Ti, and the processor model Intel Core i9 10850K. Our building and training test work is conducted under Pytorch 1.8, CUDA 11.0. The initial learning rate used in the model is 0.01; the learning-rate decay strategy is exponential decay; the weight decay is set to 0.0005; the number of training rounds is set to 100, and the batch parameters is set to 8.

3.2. Scale Priori Constraint Analysis

In this test, the width–height ratios of 2211 insulator annotation boxes were extracted from 1404 training sets of aerial images of insulators and were used as input. The annotation boxes were clustered and analyzed by controlling the cluster center number to search for the optimal width–height ratio, size, and number of anchor boxes. The width–height ratio clustering results are shown in Figure 5, where the horizontal and vertical coordinates represent the normalized width and normalized height of the insulator annotation boxes, respectively. The cross symbols in the figure represent the clustering centroids, and different clusters are distinguished by different colors. It can be seen from Figure 5 that the insulator dataset has a wide range of width–height ratio distribution, obvious differences in the width–height ratio between samples, and a large-scale span, etc. From the clustering results, it can be seen that the width–height ratio of the insulator dataset clustering center is roughly in the range of [0.3, 3].

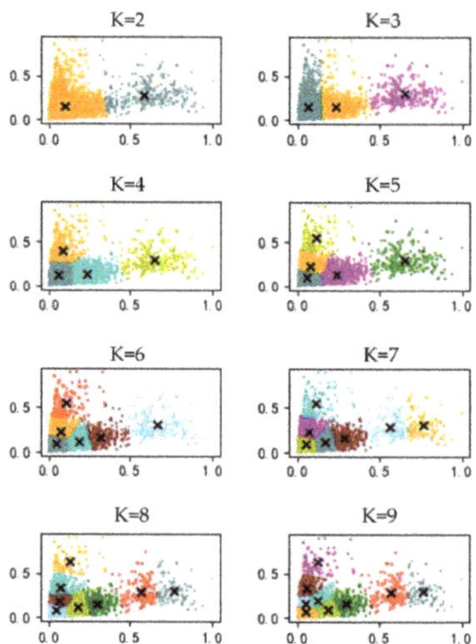

Figure 5. Aspect ratio clustering result.

Under the condition of different clustering center numbers, the clustering results of normalized width and normalized height were used as input to calculate the corresponding *CH*s, and the results are listed in Table 1.

Table 1. Clustering results under different cluster centers.

K	Center Coordinates	CH
2	[80, 134] [342, 134]	5545.599
3	[81, 90] [90, 225] [347, 133]	6018.386
4	[71, 92] [88, 226] [277, 108] [430, 167]	6624.259
5	[67, 70] [430, 167] [266, 102] [100, 282] [415, 163]	6363.596
6	[65, 68] [76, 155] [89, 272] [246, 103] [391, 130] [447, 354]	6544.842
7	[56, 70] [66, 161] [173, 97] [96, 275] [302, 117] [431, 136] [445, 375]	6325.461
8	[59, 56] [57, 126] [82, 200] [177, 95] [94, 305] [302, 118] [431, 136] [445, 375]	6313.174
9	[51, 69] [63, 156] [158, 73] [75, 278] [293, 88] [171, 184] [442, 123] [340, 179] [461, 389]	6183.416

From Table 1, it can be seen that *CH* corresponding to the number of clustering centers of four is the largest, i.e., the best clustering effect. At this time, the width–height ratio interval of [0.38, 2.52] derived from clustering is obviously beyond the preset anchor-box width–height ratio interval of [0.5, 2]. Therefore, the initial anchor-box aspect ratio is set to (1:3, 1:1, 3:1), and its corresponding aspect ratio interval [0.33, 3.0] covers the insulator aspect-ratio interval derived from the clustering, which is a good fit for the insulator size c features. Meanwhile, the center point size can be deduced from the coordinates of the center of clustering in an interval of 80^2–68^2. In order to match the insulator annotation-box size distribution, the anchor-box size is set to [2, 4, 8, 16, 32]. It can thus cover the original image size of 16^2–256^2 in the case of the minimum perceptual field and the original image size of 64^2–1024^2 in the case of the maximum perceptual field that contains the insulator annotation-box size distribution in different perceptual fields.

3.3. Ablation Test and Comparison Test

In order to evaluate the performance of the improved method, ablation tests were conducted, and the results are shown in Table 2. The YOLOv5-Orientation model was selected as the baseline model; the YOLOv5- Orientation model with the introduction of the scale priori constraints is Improved Model 1; the YOLOv5-Orientation model with the introduction of the angle priori constraints is Improved Model 2, and the RAPC-YOLO is the model proposed in this paper.

Table 2. Ablation test result.

	Scale Constraints	Angle Constraint	AP_{50}	$AP_{50\text{–}95}$
Baseline			74.35%	34.21%
Improved Model 1	√		77.92%	39.74%
Improved Model 2		√	79.23%	43.39%
RCPA-YOLO	√	√	82.33%	51.51%

As can be seen from Table 2, when GIOU was taken as 50%, the AP value of the baseline model on the tilting insulator dataset was only 74.35%. Compared to the baseline, the detection accuracies of Improved Model 1, Improved Model 2, and the RCPA-YOLO model were increased by 4.88%, 3.57%, and 7.98%, respectively. When GIOU was in the range of 50% to 95%, the AP value of the baseline model in the tilting-insulator dataset was only 34.21%. Compared with the baseline, the detection accuracy of Improved Model 1 was increased by 9.18%, and that of Improved Model 2 was increased by 5.53%. For the RCPA-YOLO model, the accuracy reached 51.51%, an increment of 17.3% compared with the baseline, a quite significant improvement.

The ablation tests show that both the scale priori constraints and the angle priori constraints can effectively improve the detection accuracy of the baseline. The priori constraint method is more effective at high GIOU thresholds, indicating that the priori constraint method can help to accurately position the tilting insulators.

The insulator detection results of the three models are demonstrated in Figure 6, where rows 1, 2, and 3 are the visualized detection results of the YOLOv5 model, the YOLOv5-Orientation model, and the RAPC-YOLO model, respectively. It can be seen from the figure that the YOLOv5 model has problems such as the misdetection of obscured insulators and the incomplete overlapping of detection bounding boxes. Besides, its detection bounding boxes are positive rectangles containing a large amount of complex background information. Although the YOLOv5-Orientation model can detect some tilting insulators, there still exist some problems such as misdetection or false detection for insulators with wide tilt angles. In contrast, our RAPC-YOLO model achieves a better detection effect for tilting insulators. The additional priori constraint module makes the oriented bounding box accurately surround the tilting insulators and realizes positioning with better precision.

Xue et al. proposed an oriented object detection model R3Det [19], based on a feature pyramid network (FPN). X. Yang et al. proposed another oriented object detection model SCRDet [20] for remotely sensed small-target object detection. These two models can be used to detect objects with arbitrary angles. In order to verify the effectiveness of the tilting-insulator detection model RAPC-YOLO, a detection performance comparison of several oriented object detection models and the RAPC-YOLO model was made in this paper under the same test conditions. The results are shown in Table 3. As can be seen, the detection accuracy of the RAPC-YOLO model in the tilting-insulator detection task is superior over other oriented object detection models. Moreover, the detection performance of the RAPC-YOLO model is more remarkable with higher intersection-over-union (IOU) thresholds, which further verifies that the priori constraint method helps to improve the detection accuracy of the model.

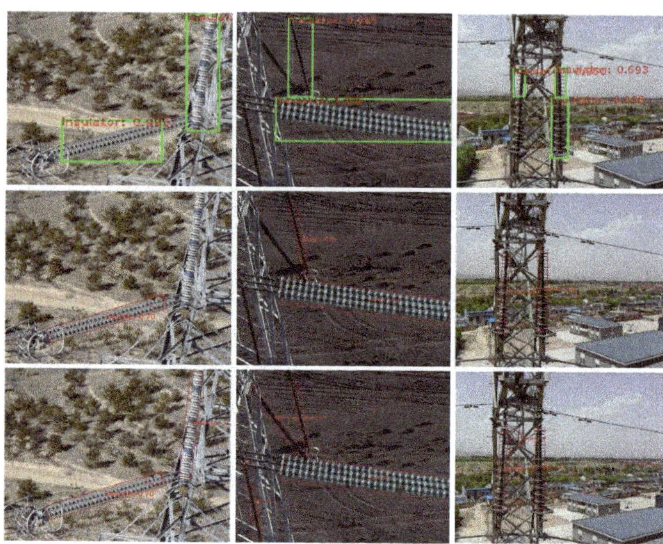

Figure 6. Comparison of detection results among our model and other detection models.

Table 3. Performance comparison of different detection models.

Model	AP$_{50}$	AP$_{50-95}$
R3Det	54.29%	14.68%
SCRDet	71.64%	35.54%
SCRDet++ [21]	73.7%	40.83%
RAPC-YOLO	82.33%	51.51%

In addition, we selectthe SCRDet++ model with the highest accuracy to carry out a visual comparative analysis with the model proposed in this paper. The results are shown in Figure 7. It can be seen from the figure that the SCRDet++ model has the problem of misdetection, especially for the shading insulator and the insulators that are relatively close to each other. In contrast, these problems are better solved in the model proposed in this paper.

Figure 7. Comparison of the detection results of our model and SCRDet++.

4. Conclusions

In this paper, we propose a RAPC-YOLO model, a new orientation detection method for tilting insulators by fusing angle regression with prior constraints. We used an oriented bounding box, angle regression loss, and rotational uncertainty function to learn the tilting features of insulators. Furthermore, we applied a clustering algorithm to learn the insulator aspect ratio and size distribution. Ablation tests and comparison tests show that our RAPC-YOLO model is an effective architecture for tilting-insulator detection tasks. In our RAPC-YOLO model, the oriented bounding box fitting the insulator edges are generated, and thus the detection effect is significantly improved compared to the baseline model, especially in the aspects of false detection and anchor-box mismatch. In addition, the results show that RAPC-YOLO is superior over other models in detection accuracy. In the future, research such as insulator-defect detection will be further carried out on the basis of the proposed RAPC-YOLO.

Author Contributions: Conceptualization, L.L. and J.Z.; methodology, L.L. and J.Z.; software, L.L.; validation, Z.C.; formal analysis, L.L. and H.F.; investigation, Y.J.; resources, L.L.; data curation, L.L.; writing—original draft preparation, J.Z.; writing—review and editing, L.L.; visualization, Z.C.; supervision, Y.J.; project administration, L.L.; funding acquisition, L.L. All authors have read and agreed to the published version of the manuscript.

Funding: This paper was supported by State Grid Hebei Electric Power Provincial Company Science and Technology Project Fund Grant Project. The sponsor is the Electric Power Research Institute of State Grid Hebei Electric Power Co., Ltd. and the project number is kj2021-016.

Institutional Review Board Statement: Not applicable.

Informed Consent Statement: Not applicable.

Data Availability Statement: Not applicable.

Acknowledgments: Special thanks are given to Electric Power Research Institute, State Grid Hebei Electric Power Co., Ltd.

Conflicts of Interest: All authors have received research grants from Electric Power Research Institute, State Grid Hebei Electric Power Co., Ltd. None of the authors has received a speaker honorarium from other companies or owns stocks in other companies. None of the authors has been involved as a consultant and expert witness in other companies. The content of the manuscript has been used in an application for patents, and none of the authors is the inventor of patents related to manuscript content.

Abbreviations

The following abbreviations are used in this manuscript:

UAV	Unmanned Aerial Vehicle
RAPC-YOLO	Regression of Angle and Prior Constraints YOLO
YOLO	You Only Look Once
GIOU	Generalized Intersection Over Union
CPLID	Chinese Power Line Insulator Dataset
AP	Average Precision
IOU	Intersection Over Union
FPN	Feature Pyramid Network

References

1. Sadykova, D.; Pernebayeva, D.; Bagheri, M.; James, A. IN-YOLO: Real-time detection of outdoor high voltage insulators using UAV imaging. *IEEE Trans. Power Deliv.* **2020**, *35*, 1599–1601. [CrossRef]
2. Zhao, W.; Xu, M.; Cheng, X.; Zhao, Z. An insulator in transmission lines recognition and fault detection model based on improved faster RCNN. *IEEE Trans. Instrum. Meas.* **2021**, *70*, 5016408. [CrossRef]
3. Tao, X.; Zhang, D.; Wang, Z.; Liu, X.; Zhang, H.; Xu, D. Detection of power line insulator defects using aerial images analyzed with convolutional neural networks. *IEEE Trans. Syst. Man Cybern. Syst.* **2020**, *50*, 1486–1498. [CrossRef]

4. Zhang, Y.; He, X.; Liu, H. An accurate and real-time method of self-blast glass insulator location based on faster R-CNN and U-net with aerial images. *CSEE J. Power Energy Syst.* **2019**, *5*, 474–482.

5. Hao, K.; Chen, G.; Zhao, L.; Li, Z.; Liu, Y.; Wang, C. An insulator defect detection model in aerial images based on Multiscale Feature Pyramid Network. *IEEE Trans. Instrum. Meas.* **2022**, *71*, 3522412. [CrossRef]

6. Mei, X.; Lu, T.; Wu, X.; Zhang, B. Insulator surface dirt image detection technology based on improved watershed algorithm. In Proceedings of the Asia-Pacific Power and Energy Engineering Conference, Shanghai, China, 27–29 March 2012; pp. 1–5. [CrossRef]

7. Zhai, Y.; Wang, D.; Zhao, Z. Insulator string location method based on airspace morphological consistency characteristics. *Chin. J. Electr. Eng.* **2017**, *37*, 1568–1578.

8. Zhang, X.; An, J.; Chen, F. A method of insulator fault detection from airborne images. In Proceedings of the 2010 Second WRI Global Congress on Intelligent Systems, Wuhan, China, 16–17 December 2010; pp. 200–203. [CrossRef]

9. Wang, Z.; Gao, Q.; Li, D.; Liu, J.; Wang, H.; Yu, X.; Wang, Y. Insulator anomaly detection method based on few-shot learning. *IEEE Access* **2021**, *9*, 94970–94980. [CrossRef]

10. Wang, C.; Wang, N.; Ho, S.-C.; Chen, X.; Song, G. Design of a new vision-based method for the bolts looseness detection in flange connections. *IEEE Trans. Ind. Electron.* **2020**, *67*, 1366–1375. [CrossRef]

11. Li, Q.; Zhao, F.; Xu, Z.; Wang, J.; Liu, K.; Qin, L. Insulator and damage detection and location based on YOLOv5. In Proceedings of the 2022 International Conference on Power Energy Systems and Applications (ICoPESA), Singapore, 25–27 February 2022; pp. 17–24.

12. Zheng, H.; Liu, Y.; Sun, Y.; Li, J.; Shi, Z.; Zhang, C.; Lai, C.S.; Lai, L.L. Arbitrary-oriented detection of insulators in thermal imagery via rotation region network. *IEEE Trans. Ind. Inform.* **2022**, *18*, 5242–5252. [CrossRef]

13. Yao, L.; Yaoyao, Q. Insulator detection dased on GIOU-YOLOv3. In Proceedings of the 2020 Chinese Automation Congress (CAC), Shanghai, China, 6–8 November 2020; pp. 5066–5071. [CrossRef]

14. Liu, C.; Yu, S.; Yu, M.; Wei, B.; Li, B.; Li, G.; Huang, W. Adaptive smooth L1 loss: A Better way to regress scene texts with extreme aspect ratios. In Proceedings of the 2021 IEEE Symposium on Computers and Communications (ISCC), Athens, Greece, 5–8 September 2021; pp. 1–7. [CrossRef]

15. Wang, Z.; Zhou, Y.; Li, G. Anomaly detection by using streaming K-means and batch K-means. In Proceedings of the 2020 5th IEEE International Conference on Big Data Analytics (ICBDA), Xiamen, China, 8–11 May 2020; pp. 11–17.

16. Wang, J.; Wang, X.; Shen, T.; Wang, Y.; Li, L.; Tian, Y.; Yu, H.; Chen, L.; Xin, J.; Wu, X.; et al. Parallel vision for long-tail regularization: Initial results from IVFC autonomous driving testing. *IEEE Trans. Intell. Veh.* **2022**, *7*, 286–299. [CrossRef]

17. Shuai, X.; Shen, Y.; Jiang, S.; Zhao, Z.; Yan, Z.; Xing, G. BalanceFL: Addressing class imbalance in long-tail federated learning. In Proceedings of the 2022 21st ACM/IEEE International Conference on Information Processing in Sensor Networks (IPSN), Milano, Italy, 4–6 May 2022; pp. 271–284.

18. Kalra, A.; Stoppi, G.; Brown, B.; Agarwal, R.; Kadambi, A. Towards rotation invariance in object detection. In Proceedings of the IEEE/CVF International Conference on Computer Vision, Montreal, BC, Canada, 11–17 October 2021; pp. 3530–3540.

19. Yang, X.; Yan, J.; Feng, Z.; He, T. R3Det: Refined single-stage detector with feature refinement for rotating object. *arXiv* **2021**, arXiv:1908.50612v6. [CrossRef]

20. Yang, X.; Yang, J.; Yan, J.; Zhang, Y.; Zhang, T.; Guo, Z.; Sun, X.; Fu, K. SCRDet: Towards more robust detection for small, cluttered and rotated objects. In Proceedings of the 2019 IEEE/CVF International Conference on Computer Vision (ICCV), Seoul, Republic of Korea, 27 October–2 November 2019; pp. 8231–8240.

21. Yang, X.; Yan, J.; Liao, W.; Yang, X.; Tang, J.; He, T. SCRDet++: Detecting small, cluttered and rotated objects via instance-level feature denoising and rotation loss smoothing. *arXiv* **2022**, arXiv:2004.13362v2. [CrossRef] [PubMed]

Article

Transmission Line Object Detection Method Based on Contextual Information Enhancement and Joint Heterogeneous Representation

Lijuan Zhao [1], Chang'an Liu [2] and Hongquan Qu [2,*]

[1] School of Control and Computer Engineering, North China Electric Power University, Beijing 102206, China
[2] School of Information, North China University of Technology, Beijing 100144, China
* Correspondence: qhqphd@ncut.edu.cn

Abstract: Transmission line inspection plays an important role in maintaining power security. In the object detection of the transmission line, the large-scale gap of the fittings is still a main and negative factor in affecting the detection accuracy. In this study, an optimized method is proposed based on the contextual information enhancement (CIE) and joint heterogeneous representation (JHR). In the high-resolution feature extraction layer of the Swin transformer, the convolution is added in the part of the self-attention calculation, which can enhance the contextual information features and improve the feature extraction ability for small objects. Moreover, in the detection head, the joint heterogeneous representations of different detection methods are combined to enhance the features of classification and localization tasks, which can improve the detection accuracy of small objects. The experimental results show that this optimized method has a good detection performance on the small-sized and obscured objects in the transmission line. The total mAP (mean average precision) of the detected objects by this optimized method is increased by 5.8%, and in particular, the AP of the normal pin is increased by 18.6%. The improvement of the accuracy of the transmission line object detection method lays a foundation for further real-time inspection.

Keywords: contextual information enhancement; joint heterogeneous representation; transmission line; object detection

Citation: Zhao, L.; Liu, C.; Qu, H. Transmission Line Object Detection Method Based on Contextual Information Enhancement and Joint Heterogeneous Representation. *Sensors* **2022**, *22*, 6855. https://doi.org/10.3390/s22186855

Academic Editor: Paweł Pławiak

Received: 14 August 2022
Accepted: 8 September 2022
Published: 10 September 2022

1. Introduction

The stable operation of the transmission line significantly affects the stability of the national economy and people's livelihood [1]. In the transmission line, there are numerous components, such as insulators, vibration dampers, pins, etc., with a small size, complex shape and various installation positions, which cause great difficulty for the traditional manual inspection [2,3]. Therefore, the unmanned aerial vehicle (UAV) with a high detection speed, efficiency and security has been widely used in transmission line inspection [4].

In the field of computer vision (e.g., image classification [5,6], object detection [7] and semantic segmentation [8,9]), CNN (convolutional neural network) is widely used as a basic framework because of its local perceptibility and translation invariance. In the last decades, lots of classical deep learning models based on CNN have been proposed, such as AlexNet, VGG [10], GoogleNet [11], ResNet [12], DenseNet [13], MobileNet [14], ShffuleNet [15], EfficientNet [16], ResNeSt [17], etc. Based on CNN, numerous methods were proposed for transmission line object detection. Li et al. [18] achieved the detection of insulator defects by adding ResNet to the backbone network of SSD. Zhao et al. [19] detected insulators under a complex background by cutting object images and using Faster R-CNN. Bao et al. [20] used the improved YOLO algorithm to detect abnormal vibration dampers on the transmission line. Tang et al. [21] detected the grading ring by changing the size of the convolution kernel in Faster R-CNN. Yang et al. [22] could identify the shockproof hammer by multi-scale fusion and depth separable convolution. The mentioned methods are aimed at the large-scale objects in a uniform background in the aerial images. However, in the transmission line, as there are many types of components with large-scale differences, complex connections, and spatial location relationships, the proportion of small-scale objects is particularly large. Almost all small-sized objects have few characteristics and are always obscured by the other large-sized object in the visual field. For the detection of small-scale objects, Zhao et al. [23] proposed a visual shape clustering network to construct a bolt defect detection model. Jiao et al. [24] proposed a context information and multi-scale pyramid network to detect bolt defects in the transmission tower. In the previous detection methods, the features of the small object were not fully extracted, and the connection and spatial location relationships among each component were also wasted, as well as the opportunity of auxiliary detection.

CNN mainly focuses on local feature extraction, but its global feature extraction capability is insufficient. Recently, transformer has gained much interest because of its advantage of a self-attention mechanism in modeling the global relationship. The self-attention mechanism is to capture the correlation of feature vectors at different spatial locations within an image. In the early exploration of the self-attention mechanism, in order to accomplish the target task, non-local operation [25] obtained the center point features by fusing the neighboring point features. DETR [26] implemented transformer-based object detection with an end-to-end format for the first time. Dosovitskiyet et al. [27] proposed visual transformer (VIT) with good performance in image classification, which could be directly applied in image block sequences. In order to replace ResNet in downstream tasks, the pyramid vision transformer (PVT) [28] was obtained by stacking multiple independent transformer encoders and introducing the pyramid structure into the transformer. Although the pure transformer has an advantage in extracting global features, the extraction of the local features is ignored. Therefore, numerous studies have begun to focus on improving the ability of global feature extraction and strengthening the modeling of local information, such as transformer in transformer [29], Swin transformer [30], Regionvit [31], etc. Transformer in transformer [29] constructs a sequence of image blocks and a sequence of super-pixels by two transformer blocks, which could achieve the encoding of the internal structure information between pixels within a patch. In order to speed up the computation, Swin transformer [30] performed a self-attention mechanism by dividing windows; at the same time, the information interaction across could be realized by shifting windows. Regionvit [31] generated the region markers and the local markers in images with different patch sizes by using region-to-local attention, and the local markers received the global information by paying attention to the region markers. In sum, the self-attention in transformer focuses on the extraction of global features and ignores the contextual information interaction between two neighboring keys.

In this study, in order to improve the detection accuracy for small objects and multi-scale objects in the transmission line, an optimized model is proposed. The optimization includes two aspects: first, in the stage of feature extraction, the advantages of the Swin transformer in global feature extraction and the convolution in local feature extraction are combined. The weight coefficients of queries and keys with rich contextual information are obtained and then applied in the value with domain information. Second, in the stage of detection, the classification and localization features of the main representation are enhanced by the heterogeneous representation of the detected object, and the advantages of different representations are fully utilized. The detection accuracy of the optimized model is examined by an experimental dataset, and the results have shown that this model has good detection accuracy, especially for small-sized transmission line objects.

2. The Detected Object

At present, the data collection of the transmission line objects is mainly carried out by the UAV. According to the technical guide of the UAV inspection for the high-voltage transmission line, in the data collection, the minimum distance between the UAV and the transmission line is 10 m, while the maximum distance is determined by the performance of the UAV and the size of the detected objects. In this study, the scenes including the grading ring and the strain clamp are selected as the detected objects. As shown in Figure 1a, the grading ring scene contains the grading ring, stay wire double plate, adjusting plate and numerous pins with a small size. As shown in Figure 1b, the strain clamp scene contains the counterweight, stay wire double plate, strain clamp, grading ring and pins.

(a)   (b)

Figure 1. The detected objects. (**a**) The grading ring scene, (**b**) the strain clamp scene.

There are many factors that play a negative effect on the transmission line object detection. In particular, the visual field proportion of the detected object to the whole image is small, especially for small-sized objects, which are always seriously obscured, such as the pins and the grading ring in the red rectangle in Figure 2. The small-sized and obscured objects in the dataset increase the difficulty of detection.

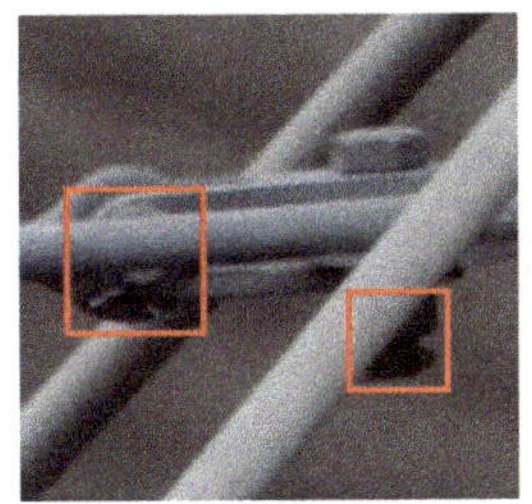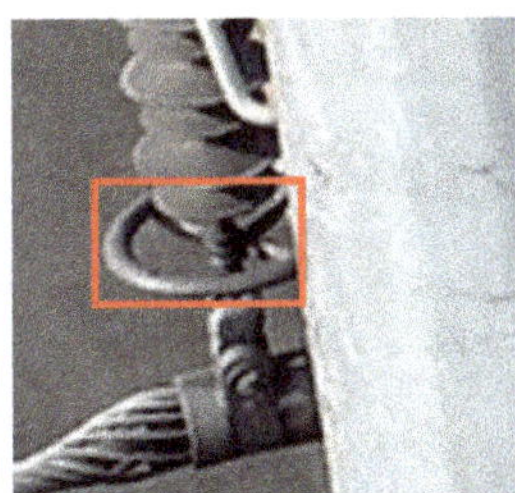

Figure 2. The small-sized and obscured fittings and connecting components.

The data in Figure 3a were obtained by a four-winged UAV with NIKON D90 and AF VR Zoom-Nikkor 80–400 mm f/4.5–5.6 D ED. Moreover, the amount and size of the various components in the transmission line are different, which causes the imbalanced category and the scale difference of the detected objects in the whole collected data. The statistics of the detected objects in the collected dataset are shown in Figure 3a. In the original dataset, the amount of the collected sample of the completed pin is the largest, followed by the counterweight, the adjusting plate, the stay wire double plate, the strain clamp, the grading ring, and the missing pin, which is the least. The size ratio of the detected object to the whole image is shown in Figure 3b. The size ratio of the grading ring is the largest, followed by the stay wire double plate, the strain clamp, the counterweight, the adjusting plate, and the pin, which is the smallest. The specific comparison is as follows: the size of the collected image is 4288 × 2848, the size range of the grading ring in the data is from 1170 to 2100, while the size of the pin ranges from 30 to 220. The size ratio of the pin is shown as a line, while the size ratio of the grading ring is shown as a block.

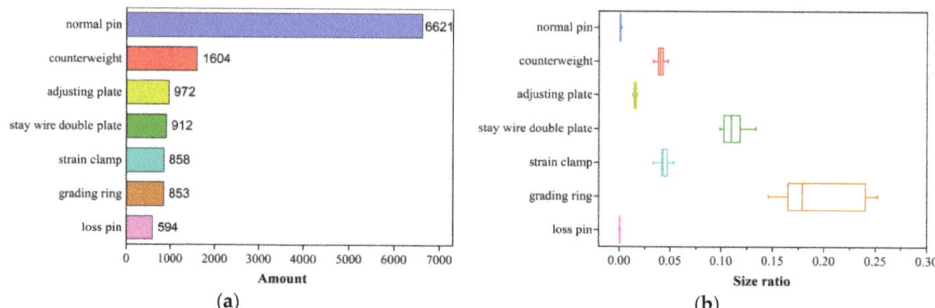

Figure 3. The statistics of the detected objects. (**a**) The amount of the detected objects, (**b**) the size ratio of the detected objects.

3. Method

In this study, we make full use of the object information in terms of both the implicit domain features and the explicit spatial location relationships in the feature extraction and the detection head, respectively. In the feature extraction, the contextual information feature is enhanced by fusing the convolution and the transformer. In the detection head, the classification and localization features are enhanced by weighting the object features of different representation types, and the heterogeneous features of different representations are fused to assist detection. The accuracy of object detection is improved from two aspects: the contextual information enhancement and the joint heterogeneous representations.

The framework of the object detection is shown in Figure 4. The Swin transformer enhanced by contextual information is used as the backbone of the network to extract features. In the process of feature fusion, FPN (feature pyramid network) [32] is used for feature fusion. In the detection head, on the basis of the original three branches of classification, regression and center-ness, a DVR (different visual representation) network is added to enhance the heterogeneous features (corner and center) of classification and localization features. The allocation of the positive and negative samples in the detection head is defined by combining the IOU of the predicted result and the anchor. In the network initialization stage, the IOU of the anchor is used a priori. In the training process, the IOU of the predicted result is added as a guide for the selection of the higher quality positive and negative samples so as to produce higher quality predictions.

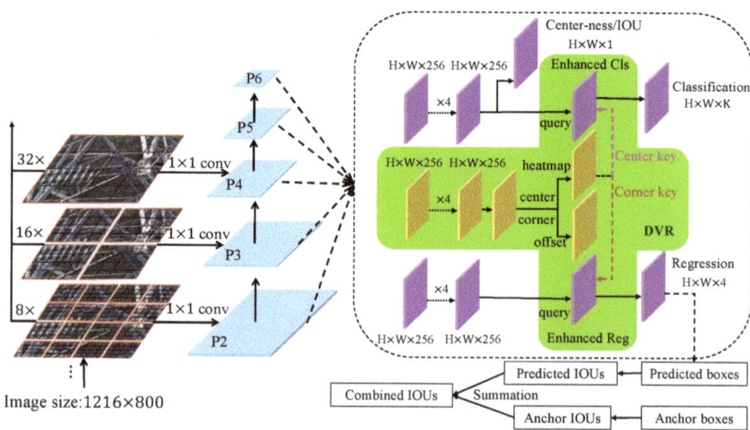

Figure 4. The framework of the heterogeneous representation auxiliary detection with contextual information. H and W are the height and width of the feature maps, respectively. K is the number of classes.

3.1. Swin Transformer Architecture with Contextual Information Enhanced

In the backbone architecture, through the fusion of the self-attention key pairs for global information extraction and the convolution for local feature information extraction, the network possesses the representation ability of the global information and the local contextual information at the same time. The properties of the convolution and the self-attention are shown in Table 1.

Table 1. The properties of the convolution and the self-attention.

Property	Convolution	Self-Attention
Translation Equivariance	✓	
Input-adaptive Weighting		✓
Local Receptive Field	✓	
Global Receptive Field		✓

Compared to the computation of CNN, the transformer is larger in computation. In this study, the input is a high-resolution image in order to reduce the computation as much as possible, and the self-attention computation of the non-overlapping windows in the Swin transformer is used as the backbone of the feature extraction. In the feature extraction, the role of the local receptive field in the shallow feature layer is much higher than that of the global receptive field, so the ability of the contextual information extraction is improved by the fusion convolution operation in the first stage. By adding the convolutional kernels into the similarity computation of key pairs and the self-attention output computation, it makes it such that the similarity computation of self-attention fuses the contextual information of neighboring ranges of key pairs, and the output global feature information contains the local receptive fields. Thus, the optimized model has a strong feature representation for small objects, and the network structure of the improved backbone architecture is shown in Figure 5.

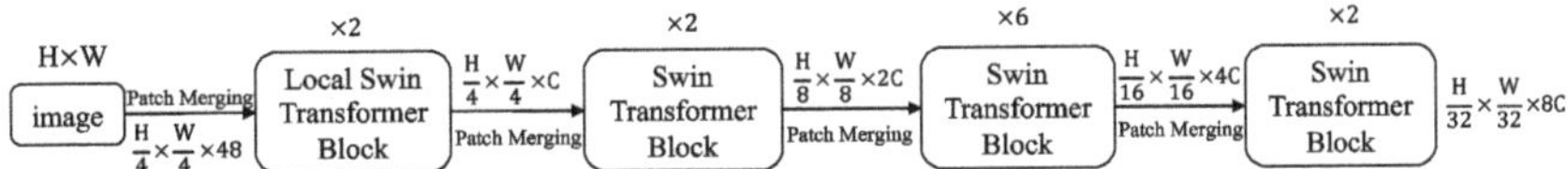

Figure 5. Swin transformer architecture with contextual information enhanced.

The convolution operation is added to the self-attention calculation inside the MHSA of the local Swin transformer block in Figure 5. The convolution and the self-attention are generalized to a unified convolutional self-attention:

$$y_i = \sum_{j\in L(i)} \alpha_{i-j}\omega_{i-j} \odot x_j \tag{1}$$

$$\alpha_{i-j} = \frac{e^{(\omega_q x_i)^T \omega_k x_j}}{\sum_{z\in L(i)} e^{(\omega_q x_i)^T \omega_k x_z}} \tag{2}$$

where $i, j \in R$ are the indexes of the spatial location, i is the index of the kernel location, and $j\in L(i)$ is the index of the local spatial neighborhood location of i. Taking a 3×3 convolution kernel as an example, i is index of the kernel location, j are the indexes included i and other eight indexes of the neighboring locations; $\alpha_{i-j}\in(0,1)$ is the weight coefficient of each location in the summation; ω_{i-j} is the projection matrix of the spatial relationship of i and j; x_j is the input feature vector; and ω_q and ω_k are the projection matrices of the query and the key, respectively.

Taking a 3×3 convolution kernel as an example, the unified convolutional self-attention model is shown in Figure 6; $\omega_{i-j} \odot x_j$ in Equation (1) is calculated by batch matrix multiplication (BMM). The generalized convolutional self-attention is expressed by a unified formula, and the calculation process of the model is adjusted by changing α_{i-j} and ω_{i-j} of the formula. In Figure 6a, the convolution kernel is applied in the input data by batch matrix multiplication, and the output, y_i, of a 3×3 matrix with the same value is obtained when the matrix α_{i-j} equals 1. In Figure 6b, through batch matrix multiplication, a dot product is obtained between the ω_v and the input data, and the output of a 3×3 matrix is obtained by combining the weight coefficients of the matrix α_{i-j}. The convolution calculation and the self-attention calculation have the advantages of the learnable filter and the dynamic kernel, respectively, which are unified by Equation (1). Therefore, the generalized convolutional self-attention incorporates the advantages of convolution and self-attention.

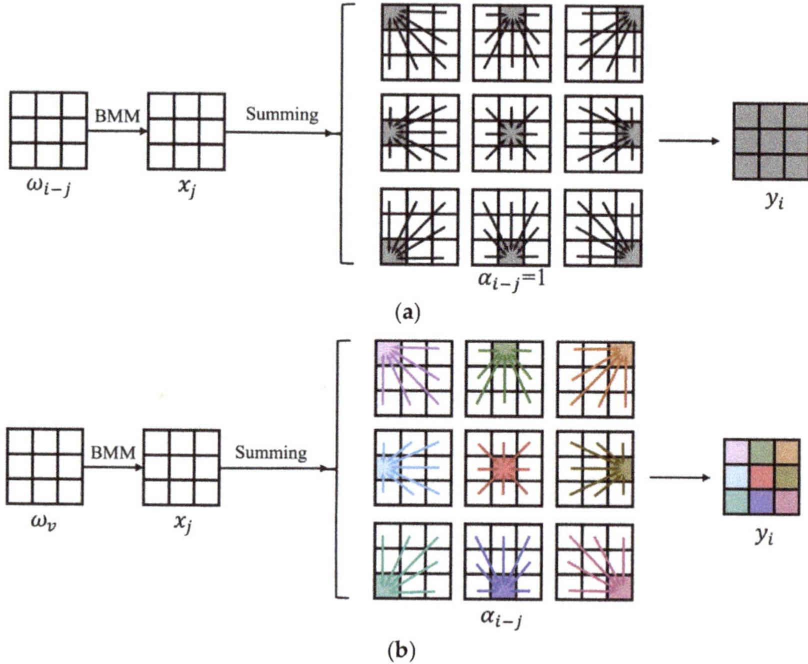

Figure 6. Illustration of the convolution and self-attention calculation. (**a**) The convolution calculation, (**b**) the self-attention calculation.

In the local Swin transformer block, the generalized convolutional self-attention is applied in the calculation process of the multi-head self-attention. The structure of the local enhanced convolutional self-attention is shown in Figure 7. The generalized convolutional self-attention is used to replace the self-attention. The weight coefficients of the query and the key with a 3×3 convolution are calculated firstly, and then, according to the weight coefficients, a weighted sum is obtained for the value that adds a 3×3 convolution, and finally, the feature fused with the contextual information is obtained.

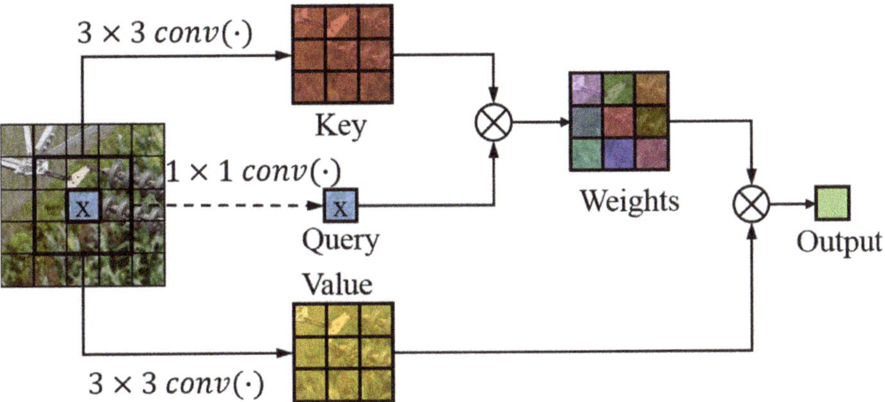

Figure 7. Contextual information enhanced convolutional self-attention module.

3.2. Heterogeneous Representation Auxiliary Detection

In the field of object detection, some methods use the bounding box as the final output, which is the dominant representation, while the other methods use corner points and center points of the bounding box as the auxiliary representation. In this study, the bounding box, center point and corner point representation are summarized in terms of heterogeneous representations. The bounding box is easier to align with annotations, the center point representation can avoid many redundant boxes from a large number of anchors and is friendly to small object detection, while the corner point representation possesses a high localization accuracy. In order to combine the advantages of the mentioned heterogeneous representations into one framework, a DVR (different visual representation) network module is added to the detection head, as shown in the green box in Figure 4. The DVR module calculates the corner and center points through a small network, and then uses the object features represented by the center and corner as the auxiliary detection to enhance the classification and localization tasks of the original network. The calculation of this module is inspired by the idea of transformer, which uses the feature of heterogeneous representation as a key to assist in enhancing the main representation feature, so as to obtain the classification and localization feature assisted by heterogeneous representation. The feature enhancement is expressed as

$$f_i^{'q} = f_i^q + \sum_j S\left(f_i^q, f_j^k, g_i^q, g_j^k\right) \cdot T_v\left(f_j^k\right) \tag{3}$$

where, $f_i^q, f_i^{'q}, g_i^q$ are the input, output and geometric vectors of the query instance i, respectively; f_j^k and g_j^k are the input and geometric vectors of the key instance j, respectively; $T_v(\cdot)$ is the value obtained by linear transformation; and $S(\cdot)$ is the similarity function calculation of i and j.

$$S\left(f_i^q, f_j^k, g_i^q, g_j^k\right) = softmax_j\left(S^A\left(f_i^q, f_j^k\right) + S^G\left(g_i^q, g_j^k\right)\right) \tag{4}$$

where $S^A\left(f_i^q, f_j^k\right)$ is the similarity in appearance between i and j; and $S^G\left(g_i^q, g_j^k\right)$ is the geometric term of the relative positions of i and j. The geometry vector of query is a 4D bounding box, and the geometry vector key is a 2D point (center point or corner point), which are obtained from the bounding box representation of the main network and the DVR network, respectively. When the dimension between the query geometry vector and the key geometry vector is different, convert the 4D bounding box into a 2D point, that is, the center point or corner point, and then calculate the similarity of the geometry vector between the key and the query.

4. Experiments

The original dataset is collected by UAV, and the useful data are selected manually. The dataset used in this study is constructed and expanded by way of rotating, changing the brightness and increasing the noise, where the preprocessing parameters are shown in Table 2. A total of 2950 images and 12,414 annotations of fittings and pins are obtained. There are seven types of objects: counterweight, grading ring, stay wire double plate, strain clamp, adjusting plate, normal pin and loss pin. The original image size is 4288×2848, and the sizes of the pin and the adjusting plate in the dataset range from 30 to 220 and from 50 to 400, respectively. The size ratio of the pin and the adjusting plate to the whole image is less than 0.12%, which is small in size, and can meet the definition of the relative size of small objects [33]. Therefore, in this study, the objects with the absolute size within 536×356 are referred to as small objects. The dataset is divided into the training set, validation set and test set, according to the amount ratio of 7:1:2. The number of classes in the three sets is not strictly at a 7:1:2 ratio. We counted the number of objects in these three sets, and the ratios of all objects are close to 7:1:2, which are listed in Table 3. The proposed model is trained and tested with a single NVIDIA GeForce RTX 2080Ti GPU, and the operating system is Ubuntu 20.04 with CUDA11.4 for training acceleration. Our experiment is implemented with PyTorch, and the batch size is set to 1 because of two aspects: first, the batch size is restricted by our hardware facilities, which only support a batch size of 1; second, the amount of the detected object in dataset is relatively small, and the convergence is good when the batch size is set as 1. The value of the epoch is proportional to the object diversity of the dataset. In this paper, the objects are less diverse, so the value of the epoch is relatively small. As shown in Figure 8, the value of the mAP of the experiment in this paper is almost a constant when the value of the epoch is larger than 39. Therefore, we set the value of the epoch as 48, which is enough to maintain the detection accuracy. We use stochastic gradient descent (SDG) for optimization. To avoid overfitting, we set the weight decay as le-4, and the learning rate as le-5.

Table 2. The preprocessing parameters of the data augmentation.

Operation Name	Description	Range of Magnitudes
Rotate	Rotate the image magnitude degrees	$[-30, 30]$
Brightness	Adjust the brightness of the image. A magnitude = 0 gives a black image, whereas magnitude = 1 gives the original image.	$[0.1, 1.9]$
Noise	Increase Gaussian noise	Mean = 0, var = 0.001

Table 3. The class ratio of each object.

Object	Training Set	Validation Set	Test Set	Class Ratio
Normal pin	4654	675	1292	7.03:1.02:1.95
Counterweight	1139	153	312	7.1:0.96:1.94
Adjusting plate	672	99	201	6.91:1.02:2.07
Stay wire double plate	648	86	178	7.11:0.94:1.95
Strain clamp	594	79	185	6.92:0.92:2.16
Grading ring	605	82	166	7.09:0.96:1.95
Loss pin	431	53	110	7.26:0.89:1.85

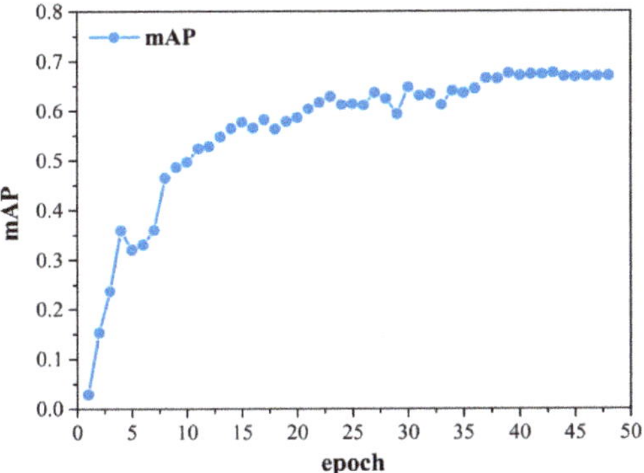

Figure 8. The mAP curve of the experiment in this paper when epoch is 48.

4.1. Experiment Results

The original dataset of this paper contains high-resolution images. In order to minimize the pixel loss in the resized image and achieve good detection performance, a large-sized image is used in the experiment, and the input size of the original dataset is set to 2332 × 1400. In order to verify the detection accuracy of the optimized method as the input image size is reduced, the experimental image sizes are selected as 2332 × 1400, 2166 × 1300, 1999 × 1200, 1666 × 1000, and 1333 × 800. The corresponding learning times for these datasets are about 20 h, 18 h, 16 h, 13 h, and 10 h, respectively. The detection results are shown in Table 4; it can be found that when the image size increases from 1333 × 800 to 1666 × 1000, the total mAP is increased by 6.8%, and the AP of the normal pin is increased by 12.3%. When the image size increases from 1666 × 1000 to 1999 × 1200, the total mAP is increased by 7.8%, and the AP of the normal pin is increased by 14.3%. When the image size increases to 1999 × 1200, the increase in the mAP is small and is less than 1%. When the image size increases to 2332 × 1400 and 2166 × 1300, the total mAP is barely increased, and only the APs of the normal pin are increased by 2.6% and 1.2%, respectively. In general, with the increase in the input image size, the average detection accuracy of all objects is increased, especially for the small object of the normal pin.

Table 4. Comparison of the detected objects with different image sizes.

Input Image Size	Normal Pin	Counterweight	Grading Ring	Stay Wire Double Plate	Strain Clamp	Loss Pin	Adjusting Plate	mAP	Para/MB	FLOPs
1333 × 800	39.3	58.3	73.1	61.1	64.7	23.8	47.3	52.5	4126	185.03
1666 × 1000	51.6	65.1	77.2	65.2	72.3	29.6	54.1	59.3	5387	304.03
1999 × 1200	65.9	72.7	83.9	69.1	79	36.2	62.6	67.1	6644	425.24
2166 ×1300	68.5	72.8	84.3	69.2	79.3	36.5	62.7	67.6	7677	660.08
2332 × 1400	69.7	72.9	84.5	69.4	79.7	36.9	62.9	68	8534	782.54

Figure 9 shows the visual results of the detected objects with different image sizes. Compared with Figure 9a–c, it can be found that with the increase in the input image size, the number of the detected normal pins on the parallel hanging plate (in the enlarged red circle) is increased, while the number of missed detections is decreased from 2 to 0. In Figure 9d,e, the obscured normal pin on the adjustment plate also can be detected. In Figure 9f, with the lowest image size, there is a false detection of the normal pin in the enlarged red circle. In Figure 9g, a missed detection of the normal pin located on the stay wire double plate exists. In Figure 9h,i, a normal pin at the same position is detected, and in Figure 9j, two adjacent normal pins at this position are accurately detected. Therefore, when the input image size is increased, the small and obscured objects can be successfully detected, and the detection performance is better.

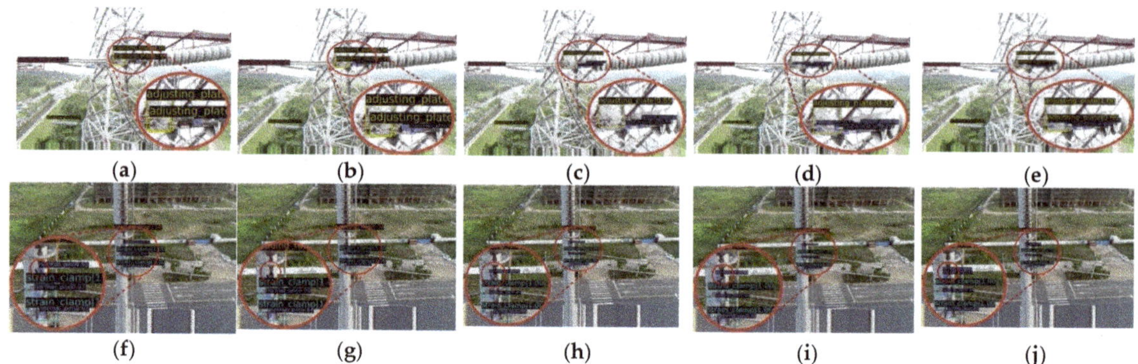

Figure 9. Visual results of the detected objects with different image sizes. (**a**) Size with 1333 × 800, (**b**) size with 1666 × 1000, (**c**) size with 1999 × 1200, (**d**) size with 2166 × 1300, (**e**) size with 2332 × 1400, (**f**) size with 1333 × 800, (**g**) size with 1666 × 1000, (**h**) size with 1999 × 1200, (**i**) size with 2166 × 1300, (**j**) size with 2332 × 1400.

In sum, with the increase in the input image size, the detection accuracy is increased, but it is worth noting that the number of parameters (Para, measuring the spatial complexity) and the floating points of operations (FLOPs, measuring the computational complexity) are increased at the same time. Therefore, 1999 × 1200 is selected as the input image size in this study, which is based on the consideration of the calculation cost and the detection accuracy affected by the pixel loss.

The baseline method, dynamic ATSS, is compared with the optimized method based on the contextual information enhancement and joint heterogeneous representation (CIE-JHR) in this study. The comparison of the AP of single-type fittings are shown in Figure 10. Compared to the dynamic ATSS, the AP of the most fittings are improved by the CIE-JHR method, except for the adjusting plate object. The greatest increase in the AP is 18.6% for the normal pin, and the smallest increase in the AP is 1.5% for the loss pin. The AP of the counterweight, grading ring, stay wire double plate and strain clamp are increased by 2%, 2%, 3% and 4%, respectively. For the adjusting plate and loss pin, the AP is barely increased or is even decreased. The reasons are that the pin and the adjusting plate are small in size, and the samples of the loss pin and adjusting plate in the dataset are small. In addition, the loss pins are also easily confused with the normal pins, and the side view morphology of the adjustment plate is easily confused with the parallel hanging plate. In sum, these experimental results show that the optimized method can effectively improve the detection accuracy of almost all objects in the case of sufficient samples.

There are numerous methods for the detection of fittings and pins, such as SSD, Faster R-CNN, Retina Net, Cascade R-CNN, FCOS, and Dynamic ATSS. These methods with the same hyper-parameters are compared with the optimized method (CIE-JHR) in this study; AP50 (when IOU threshold is 0.5), Para and Time are used as the evaluation indicators. Para represents the number of model parameters in the training stage, and Time represents the detection time per image in the test stage. As shown in Table 5, the total mAP of the detected objects by CIE-JHR method is the largest, and the next are the Dynamic ATSS, FCOS, Cascade R-CNN, Retina Net, Faster R-CNN, and that by SSD is the smallest. Therefore, the CIE-JHR method possesses the best detection performance. For the small-sized objects, such as normal pin and loss pin, compared to the SSD method, the AP of the former object and the latter object detected by CIE-JHR method is increased significantly (62.8% and 41.7%, respectively). For the adjusting plate, the AP of CIE-JHR method is barely increased compared to the others methods, and is slightly decreased compared to the Dynamic ATSS method, the main reasons for which are explained in the preceding paragraph.

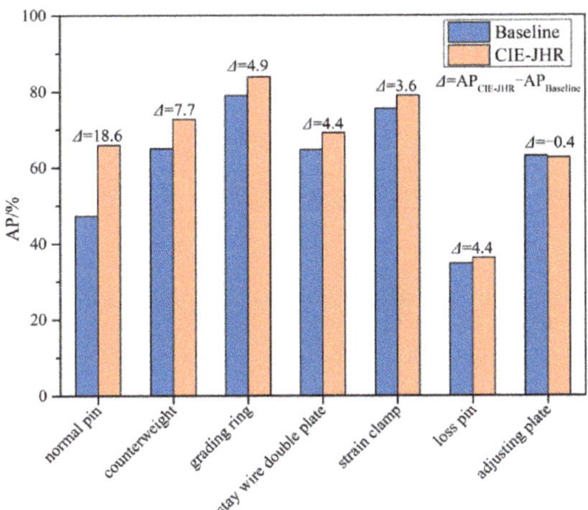

Figure 10. Comparison of the detected objects by CIE-JHR and dynamic ATSS methods.

Table 5. Comparison of the detection results.

Methods	Normal Pin	Counterweight	Grading Ring	Stay Wire Double Plate	Strain Clamp	Loss Pin	Adjusting Plate	mAP	Para /MB	Time /ms
SSD512 [34]	24.5	54.4	68.2	57.1	63.5	21.1	49.5	48.3	1754	87
Faster R-CNN [35]	41.5	60.1	73.2	58.2	69.3	25.9	56.2	54.9	2452	114
Retina Net [36]	43.5	61.4	76.2	61.1	71.5	31.6	60.5	58	2835	128
Cascade R-CNN [37]	45.7	64.8	77.1	62.7	72.6	31.5	61.4	59.4	5742	250
FCOS [38]	46.5	63.6	78	62.4	73.6	32.9	62.1	60	4372	107
Dynamic ATSS [39]	47.3	65	79	64.7	75.4	34.7	63	61.3	4642	120
CIE-JHR	65.9	72.7	83.9	69.1	79	36.2	62.6	67.1	6644	142

The visual comparison of some typical object detection by the Dynamic ATSS method and CIE-JHR method are shown in Figure 11. Compared with Figure 11a,e, it can be found that the CIE-JHR method successfully detects the normal pin located on the adjustment plate, while the dynamic ATSS algorithm does not detect the normal pin. Comparison between Figure 11b,f show that the CIE-JHR method successfully detects the loss pin located on the strain clamp and excludes the mis-detected normal pin located on the front side of the stay wire double plate. Comparison between Figure 11c,j shows that the CIE-JHR method detects some small adjusting plates and counterweight, but there is a missed detection for the normal pin located on the stay wire double plate. Compared with Figure 11d,h, the CIE-JHR method detects the loss pin located on the strain clamp, while the Dynamic ATSS method fails to detect the loss pin. In summary, the CIE-JHR method has a good detection performance on small-sized objects, especially for the pin located on the adjusting plate, stay wire double plate and strain clamp. Therefore, the accuracy of object detection is successfully improved by adding the contextual information enhancement and the joint heterogeneous representation.

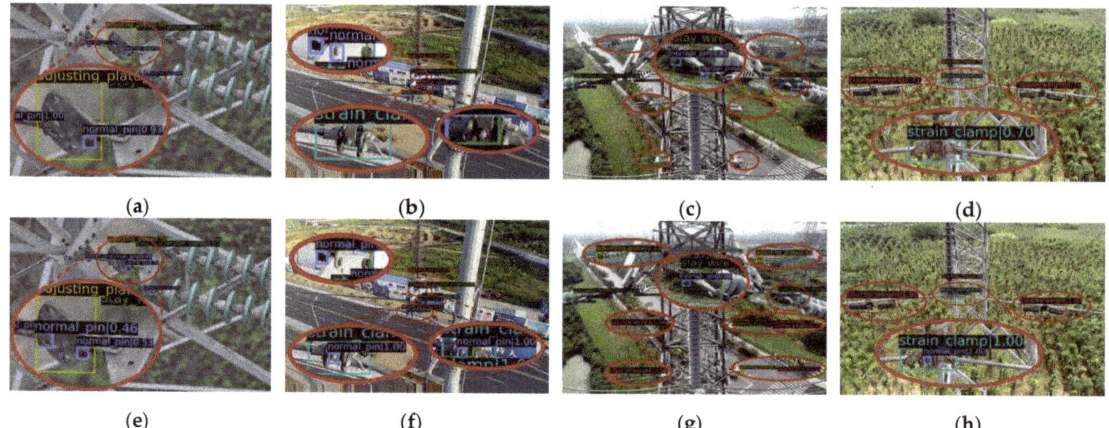

Figure 11. Visual results of the detected object by Dynamic ATSS and CIE-JHR method. (**a**) Insulator string and tower connection based on Dynamic ATSS algorithm, (**b**) strain clamp scene based on Dynamic ATSS algorithm, (**c**) multiple insulator string and tower connection based on Dynamic ATSS algorithm, (**d**) strain clamp and counterweight scene based on Dynamic ATSS algorithm, (**e**) insulator string and tower connection based on CIE-JHR method, (**f**) strain clamp scene based on CIE-JHR method, (**g**) multiple insulator string and tower connection based on CIE-JHR method, (**h**) strain clamp and counterweight scene based on CIE-JHR method.

4.2. Ablation Analysis

In order to prove the effectiveness of the contextual information enhancement (CIE) of the feature extraction module and the joint heterogeneous representation (JHR) in the detection stage, ablation experiments are carried out based on the CIE-JHR method and the Dynamic ATSS method; the results are shown in Table 6. Compared to the pure Dynamic ATSS method, the AP, AP50 and AP75 of the Dynamic ATSS method with the CIE module are increased by 2.3%, 3.6% and 2.1%, respectively, which indicates that the feature extraction ability with CIE is significantly improved. The AP, AP50 and AP75 of the Dynamic ATSS method with the JHR module are increased by 1.8%, 3.5% and 2.7%, respectively, which indicates that the combination of the heterogeneous representations is conducive to the improvement of the detection accuracy. The AP, AP50 and AP75 of the CIE-JHR method are increased by 6.7%, 5.8% and 6.7%, respectively. These results show that the optimization of the contextual information enhancement in the feature extraction stage and the joint heterogeneous representation in the object detection stage can effectively improve the accuracy of the object detection.

Table 6. The results of ablation experiments.

Method	Backbone	Head	AP	AP50	AP75
Baseline			35.6	61.3	37.4
	√		37.9	64.9	39.5
		√	37.4	64.8	40.1
CIE-JHR	√	√	42.3	67.1	44.1

5. Conclusions

In this study, a transmission line object detection method is proposed, which combines the contextual information and the joint heterogeneous representation. In this method, the Swin transformer is used as the backbone of the network structure, and the convolution is added in the self-attention calculation of the low-level structure, which can make full use of the contextual information. In the detection stage, the classification and regression tasks are carried out by combining heterogeneous representation features, which could improve the detection accuracy. The experimental results show that the total mAP of the detected objects by the optimized method is 67.1%, which is increased by 5.8% compared to the baseline algorithm, especially, the AP of the small-sized normal pins is 65.9%, which is increased by 18.6% compared to the baseline algorithm. It has a good detection effect for small objects obscured by other components and adjacent small objects with close distance. The optimized method can effectively improve the average accuracy of the transmission line object detection, which is beneficial to successfully detect the potential hazard of the small-sized and obscured objects in operation, and also can help to maintain the stable and safe operation of the transmission line. Moreover, the increase in the detection accuracy of the small-sized objects is also beneficial to the establishment of the efficient real-time detection of the transmission line object. In further study, the complex relationship between the transmission line hardware can be considered, and the artificial prior knowledge can be used to improve the object detection performance.

Author Contributions: Conceptualization, L.Z.; methodology, L.Z.; software, L.Z.; validation, L.Z.; formal analysis, L.Z.; investigation, H.Q.; resources, C.L.; data curation, L.Z.; writing—original draft preparation, L.Z.; writing—review and editing, C.L.; visualization, L.Z.; supervision, C.L.; project administration, C.L.; funding acquisition, H.Q. All authors have read and agreed to the published version of the manuscript.

Funding: This research was funded by National Key R&D Program of China (No. 2020YFB1600704).

Institutional Review Board Statement: Not applicable.

Informed Consent Statement: Not applicable.

Data Availability Statement: Not applicable.

Conflicts of Interest: The authors declare no conflict of interest.

References

1. Park, K.-C.; Motai, Y.; Yoon, J.R. Acoustic Fault Detection Technique for High-Power Insulators. *IEEE Trans. Ind. Electron.* **2017**, *64*, 9699–9708. [CrossRef]
2. Liang, H.; Zuo, C.; Wei, W. Detection and Evaluation Method of Transmission Line Defects Based on Deep Learning. *IEEE Access* **2020**, *8*, 38448–38458. [CrossRef]
3. Sadykova, D.; Pernebayeva, D.; Bagheri, M.; James, A. IN-YOLO: Real-Time Detection of Outdoor High Voltage Insulators Using UAV Imaging. *IEEE Trans. Power Deliv.* **2020**, *35*, 1599–1601. [CrossRef]
4. Meng, L.; Peng, Z.; Zhou, J.; Zhang, J.; Lu, Z.; Baumann, A.; Du, Y. Real-Time Detection of Ground Objects Based on Unmanned Aerial Vehicle Remote Sensing with Deep Learning: Application in Excavator Detection for Pipeline Safety. *Remote Sens.* **2020**, *12*, 182. [CrossRef]
5. Wambugu, N.; Chen, Y.; Xiao, Z.; Tan, K.; Wei, M.; Liu, X.; Li, J. Hyperspectral Image Classification on Insufficient-Sample and Feature Learning Using Deep Neural Networks: A Review. *Int. J. Appl. Earth Obs. Geoinf.* **2021**, *105*, 102603. [CrossRef]
6. Wu, Y.; Mu, G.; Qin, C.; Miao, Q.; Ma, W.; Zhang, X. Semi-Supervised Hyperspectral Image Classification via Spatial-Regulated Self-Training. *Remote Sens.* **2020**, *12*, 159. [CrossRef]
7. Zou, Z.; Shi, Z.; Guo, Y.; Ye, J. Object Detection in 20 Years: A Survey. *arXiv* **2019**, arXiv:1905.05055.
8. Yuan, X.; Shi, J.; Gu, L. A Review of Deep Learning Methods for Semantic Segmentation of Remote Sensing Imagery. *Expert Syst. Appl.* **2021**, *169*, 114417. [CrossRef]
9. Wu, Y.; Li, J.; Yuan, Y.; Qin, A.K.; Miao, Q.-G.; Gong, M.-G. Commonality Autoencoder: Learning Common Features for Change Detection From Heterogeneous Images. *IEEE Trans. Neural Netw. Learn. Syst.* **2021**, *33*, 4257–4270. [CrossRef]
10. Simonyan, K.; Zisserman, A. Very Deep Convolutional Networks for Large-Scale Image Recognition. *arXiv* **2015**, arXiv:1409.1556.
11. Szegedy, C.; Liu, W.; Jia, Y.; Sermanet, P.; Reed, S.; Anguelov, D.; Erhan, D.; Vanhoucke, V.; Rabinovich, A. Going Deeper With Convolutions. In Proceedings of the IEEE Conference on Computer Vision and Pattern Recognition, Boston, MA, USA, 7–12 June 2015; pp. 1–9.
12. He, K.; Zhang, X.; Ren, S.; Sun, J. Deep Residual Learning for Image Recognition. In Proceedings of the IEEE Conference on Computer Vision and Pattern Recognition, Las Vegas, NV, USA, 27–30 June 2016; pp. 770–778.
13. Huang, G.; Liu, Z.; van der Maaten, L.; Weinberger, K.Q. Densely Connected Convolutional Networks. In Proceedings of the IEEE Conference on Computer Vision and Pattern Recognition, Honolulu, HI, USA, 21–26 July 2017; pp. 4700–4708.

14. Howard, A.G.; Zhu, M.; Chen, B.; Kalenichenko, D.; Wang, W.; Weyand, T.; Andreetto, M.; Adam, H. MobileNets: Efficient Convolutional Neural Networks for Mobile Vision Applications. *arXiv* **2017**, arXiv:1704.04861.
15. Zhang, X.; Zhou, X.; Lin, M.; Sun, J. ShuffleNet: An Extremely Efficient Convolutional Neural Network for Mobile Devices. In Proceedings of the IEEE Conference on Computer Vision and Pattern Recognition, Honolulu, HI, USA, 21–26 July 2017; pp. 6848–6856.
16. Tan, M.; Le, Q. EfficientNet: Rethinking Model Scaling for Convolutional Neural Networks. In Proceedings of the 36th International Conference on Machine Learning, Long Beach, CA, USA, 9–15 June 2019; pp. 6105–6114.
17. Zhang, H.; Wu, C.; Zhang, Z.; Zhu, Y.; Lin, H.; Zhang, Z.; Sun, Y.; He, T.; Mueller, J.; Manmatha, R.; et al. ResNeSt: Split-Attention Networks. In Proceedings of the IEEE/CVF Conference on Computer Vision and Pattern Recognition, New Orleans, LA, USA, 19–20 June 2022; pp. 2736–2746.
18. Li, W.; Zheng, W.; Wang, N.; Zhao, H. Research on Detection Method of Insulator Defects on Transmission Lines Based on SSD Algorithm. *Instrum. Cust.* **2019**, *26*, 1–4.
19. Zhao, W.; Xu, M.; Cheng, X.; Zhao, Z. An Insulator in Transmission Lines Recognition and Fault Detection Model Based on Improved Faster RCNN. *IEEE Trans. Instrum. Meas.* **2021**, *70*, 1–8. [CrossRef]
20. Bao, W.; Ren, Y.; Wang, N.; Hu, G.; Yang, X. Detection of Abnormal Vibration Dampers on Transmission Lines in UAV Remote Sensing Images with PMA-YOLO. *Remote Sens.* **2021**, *13*, 4134. [CrossRef]
21. Tang, Y.; Han, J.; Wei, W.; Ding, J.; Peng, X. Research on Part Recognition and Defect Detection of Transmission Line in Deep Learning. *Electron. Meas. Technol.* **2018**, *41*, 60–65.
22. Yang, G.; Sun, C.; Zhang, N.; Jin, T.; Xu, C.; Wu, T.; Zhang, X. Detection of Key Components of Transmission Lines Based on Multi-Scale Feature Fusion. *Electr. Meas. Instrum.* **2020**, *57*, 54–59.
23. Zhao, Z.; Qi, H.; Qi, Y.; Zhang, K.; Zhai, Y.; Zhao, W. Detection Method Based on Automatic Visual Shape Clustering for Pin-Missing Defect in Transmission Lines. *IEEE Trans. Instrum. Meas.* **2020**, *69*, 6080–6091. [CrossRef]
24. Jiao, R.; Liu, Y.; He, H.; Ma, X.; Li, Z. A Deep Learning Model for Small-Size Defective Components Detection in Power Transmission Tower. *IEEE Trans. Power Deliv.* **2022**, *37*, 2551–2561. [CrossRef]
25. Wang, X.; Girshick, R.; Gupta, A.; He, K. Non-Local Neural Networks. *arXiv* **2018**, arXiv:1711.07971.
26. Carion, N.; Massa, F.; Synnaeve, G.; Usunier, N.; Kirillov, A.; Zagoruyko, S. End-to-End Object Detection with Transformers. In *Computer Vision—ECCV 2020*; Vedaldi, A., Bischof, H., Brox, T., Frahm, J.-M., Eds.; Lecture Notes in Computer Science; Springer International Publishing: Cham, Switzerland, 2020; pp. 213–229. [CrossRef]
27. Dosovitskiy, A.; Beyer, L.; Kolesnikov, A.; Weissenborn, D.; Zhai, X.; Unterthiner, T.; Dehghani, M.; Minderer, M.; Heigold, G.; Gelly, S.; et al. An Image Is Worth 16x16 Words: Transformers for Image Recognition at Scale. *arXiv* **2021**, arXiv:2010.11929.
28. Wang, W.; Xie, E.; Li, X.; Fan, D.-P.; Song, K.; Liang, D.; Lu, T.; Luo, P.; Shao, L. Pyramid Vision Transformer: A Versatile Backbone for Dense Prediction Without Convolutions. In Proceedings of the IEEE/CVF International Conference on Computer Vision, Montreal, QC, Canada, 10–17 October 2021; pp. 568–578.
29. Han, K.; Xiao, A.; Wu, E.; Guo, J.; XU, C.; Wang, Y. Transformer in Transformer. In *Advances in Neural Information Processing Systems*; Curran Associates, Inc.: Red Hook, NY, USA, 2021; Volume 34, pp. 15908–15919.
30. Liu, Z.; Lin, Y.; Cao, Y.; Hu, H.; Wei, Y.; Zhang, Z.; Lin, S.; Guo, B. Swin Transformer: Hierarchical Vision Transformer Using Shifted Windows. In Proceedings of the IEEE/CVF International Conference on Computer Vision, Montreal, QC, Canada, 10–17 October 2021; pp. 10012–10022.
31. Chen, C.-F.; Panda, R.; Fan, Q. RegionViT: Regional-to-Local Attention for Vision Transformers. *arXiv* **2022**, arXiv:2106.02689.
32. Lin, T.-Y.; Dollar, P.; Girshick, R.; He, K.; Hariharan, B.; Belongie, S. Feature Pyramid Networks for Object Detection. In Proceedings of the IEEE Conference on Computer Vision and Pattern Recognition, Honolulu, HI, USA, 21–26 July 2017; pp. 2117–2125.
33. Lin, T.-Y.; Maire, M.; Belongie, S.; Hays, J.; Perona, P.; Ramanan, D.; Dollár, P.; Zitnick, C.L. Microsoft COCO: Common Objects in Context. In *Computer Vision—ECCV 2014*; Fleet, D., Pajdla, T., Schiele, B., Tuytelaars, T., Eds.; Lecture Notes in Computer Science; Springer International Publishing: Cham, Switzerland, 2014; pp. 740–755. [CrossRef]
34. Liu, W.; Anguelov, D.; Erhan, D.; Szegedy, C.; Reed, S.; Fu, C.-Y.; Berg, A.C. SSD: Single Shot MultiBox Detector. In *Computer Vision—ECCV 2016*; Leibe, B., Matas, J., Sebe, N., Welling, M., Eds.; Lecture Notes in Computer Science; Springer International Publishing: Cham, Switzerland, 2016; pp. 21–37. [CrossRef]
35. Ren, S.; He, K.; Girshick, R.; Sun, J. Faster R-CNN: Towards Real-Time Object Detection with Region Proposal Networks. In *Advances in Neural Information Processing Systems*; Curran Associates, Inc.: Red Hook, NY, USA, 2015; Volume 28.
36. Lin, T.-Y.; Goyal, P.; Girshick, R.; He, K.; Dollar, P. Focal Loss for Dense Object Detection. In Proceedings of the IEEE International Conference on Computer Vision, Venice, Italy, 22–29 October 2017; pp. 2980–2988.
37. Cai, Z.; Vasconcelos, N. Cascade R-CNN: Delving Into High Quality Object Detection. In Proceedings of the IEEE/CVF Conference on Computer Vision and Pattern Recognition, Salt Lake, UT, USA, 18–23 June 2018; pp. 6154–6162.
38. Tian, Z.; Shen, C.; Chen, H.; He, T. FCOS: Fully Convolutional One-Stage Object Detection. In Proceedings of the IEEE/CVF International Conference on Computer Vision, Seoul, Korea, 27 October–02 November 2019; pp. 9627–9636.
39. Zhang, T.; Luo, B.; Sharda, A.; Wang, G. Dynamic Label Assignment for Object Detection by Combining Predicted IoUs and Anchor IoUs. *J. Imaging* **2022**, *8*, 193. [CrossRef] [PubMed]

 *sensors*

Article

Improved Algorithm for Insulator and Its Defect Detection Based on YOLOX

Gujing Han [1], Tao Li [1,*], Qiang Li [2], Feng Zhao [2], Min Zhang [1], Ruijie Wang [1], Qiwei Yuan [1], Kaipei Liu [3] and Liang Qin [3]

[1]   Department of Electronic and Electrical Engineering, Wuhan Textile University, Wuhan 430200, China
[2]   State Grid Information & Telecommunication Group Co., Ltd., Beijing 102211, China
[3]   School of Electrical and Automation, Wuhan University, Wuhan 430072, China
*   Correspondence: ltao0301@163.com

Abstract: Aerial insulator defect images have some features. For instance, the complex background and small target of defects would make it difficult to detect insulator defects quickly and accurately. To solve the problem of low accuracy of insulator defect detection, this paper concerns the shortcomings of IoU and the sensitivity of small targets to the model regression accuracy. An improved SIoU loss function was proposed based on the regular influence of regression direction on the accuracy. This loss function can accelerate the convergence of the model and make it achieve better results in regressions. For complex backgrounds, ECA (Efficient Channel Attention Module) is embedded between the backbone and the feature fusion layer of the model to reduce the influence of redundant features on the detection accuracy and make progress in the aspect. As a result, these experiments show that the improved model achieved 97.18% mAP which is 2.74% higher than before, and the detection speed could reach 71 fps. To some extent, it can detect insulator and its defects accurately and in real-time.

Keywords: aerial insulator images; object detection; YOLOX; small target; SIoU

Citation: Han, G.; Li, T.; Li, Q.; Zhao, F.; Zhang, M.; Wang, R.; Yuan, Q.; Liu, K.; Qin, L. Improved Algorithm for Insulator and Its Defect Detection Based on YOLOX. *Sensors* **2022**, *22*, 6186. https://doi.org/10.3390/s22166186

Academic Editors: Ke Zhang and Yincheng Qi

Received: 25 July 2022
Accepted: 17 August 2022
Published: 18 August 2022

1. Introduction

Insulators are key components that provide electrical insulation and mechanical support for current-carrying conductors on high-voltage transmission lines. Defects are likely to occur due to various factors such as transient loads, mechanical stress, atmospheric conditions, etc. Furthermore, they might then threaten the stable operation of transmission lines which highly impacts the security of the power system. A UAV (Unmanned Aerial Vehicle) is more efficient and convenient as it offers visual assessments of structures. Therefore, it has gradually replaced the traditional manual inspection method. The detection of insulator defects based on aerial images consequently has become popular. However, insulator defects in aerial images often exhibit small targets and complex backgrounds in the dataset. Therefore, it is difficult to detect insulator defects quickly and accurately.

Traditional methods of insulator defect detection focus on color, texture, edge, and other features [1–4]. This kind of method relies on high-quality images and appropriate shooting angles. It might suffer from weak robustness.

Object detection algorithms based on deep learning have been widely used in power systems due to the good performance of the generalization capability and the ability to extract features from complex backgrounds [5,6]. They are generally divided into two categories: one-stage algorithm and two-stage algorithm. These two-stage object detection algorithms mostly use RPN (Region Proposal Network) to reduce the interference of complex background on insulator and its defect detection. Although the precision of insulator detection could be improved, the low efficiency and slow speed cannot be ignored [7–15].

The YOLO series is a typical one-stage object detection algorithm, which eliminates the RPN and generates the position coordinates and category probability of the object through a single detection, which can quickly and accurately complete the detection task [16]. Since its inception, the YOLO series of the algorithm is gradually developed in terms of accuracy and speed. Moreover, it has evolved a variety of more advantageous algorithm models such as YOLOv3, YOLOv4, YOLOv5, and YOLOX. In terms of the application in defect detection, Wang et al. [17] used Gaussian parameters to model the coordinates of the predicted box, which improved the accuracy of the YOLOv3 algorithm to detect defects to some extent. Zhang et al. [18] used the YOLOv3 algorithm with a dense FPN structure to improve the utilization of deep semantic information and shallow localization information, reduce the number of model parameters, and improve the detection accuracy of insulator defects. Shen et al. [19] and Duan et al. [20] suggested that the defect detection accuracy of the YOLOv3 algorithm could be improved by optimizing the regression loss. Tang et al. [21] divided the task of defect detection into two parts and improved the accuracy by using YOLOv4 to detect defects in insulators segmented by U-net. Lv et al. [22] investigated the effect of clustering algorithms on the detection results of the model, studied the effect of regression loss on targets at different scales, and proposed an intelligent identification method for electrical devices based on the YOLOv4 algorithm. Qiu et al. [23] used depthwise separable convolution to reduce the number of parameters of the YOLOv4 algorithm, improved the detection speed of the model, and used the Laplace sharpening method to preprocess the insulator image. It actually alleviated the problem of reduced detection accuracy caused by model lightweighting. Moreover, these studies are mainly based on YOLOv4 or earlier versions, and there are few studies about the regression perspective of regression loss on model accuracy.

Compared with the above algorithms, YOLOX has a faster detection speed and higher accuracy on the COCO dataset [24]. The detection speed of its lightweight model YOLOX-S achieved 75 fps which could make progress in the speed of defect detection. However, this model still suffers from the problem of low accuracy when it detects defects. At the same time, its detection results are vulnerable to the influence of complex backgrounds.

To improve the efficiency and accuracy of detection of insulator defects in high-voltage transmission lines, this paper discusses the problem that defects are difficult to detect and optimizes the YOLOX algorithm by further researching the regression loss. As for complex backgrounds, the influence of the attention mechanism on the model accuracy is also considered.

This paper proceeds as follows: It researches and analyses the defects of the regression loss of the model, considering the shortcomings of this regression loss function and further studies the law of the influence of regression angle on the accuracy of the model. The effects of different attention mechanisms on the detection effect of the model are analyzed. An improved YOLOX-S-based insulator defect detection method is proposed to achieve better results without almost changing detection speed.

2. Structure and Characteristics of the YOLOX-S Model

The structure of the YOLOX-S model is shown in Figure 1, which can be divided into three parts: Backbone (feature extraction network), Neck (feature fusion network), and Head (prediction network).

The Backbone including CSPDarknet performs convolution calculation on the input image, extracts sample features, and generates five feature layers containing different levels of semantic information. Finally, it selects the last three feature layers as the input information of the Neck.

The Neck part uses PANet (Path Aggregation Network) to fuse the feature information extracted by the Backbone part. Therefore, it not only contains the information of position, texture, edge, and others in low layers but also the strong semantic information in high layers.

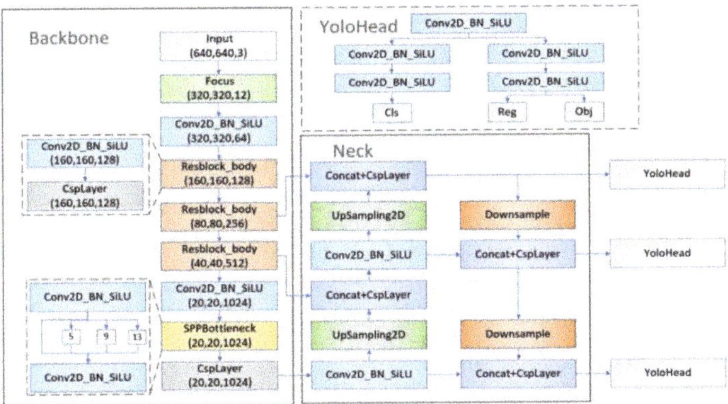

Figure 1. Model structure of YOLOX.

Different from the Head of the previous models of the YOLO series, the YOLOX model decoupled the classification and the localization task. it solved the conflicts caused by the coupled two tasks and improved the convergence speed.

Since YOLOX adopts the idea of Anchor-free, it is unnecessary to preset anchor box which greatly reduces the computation of anchor-box clustering and improves the detection speed of the model. However, the model cannot effectively mitigate the influence of redundant information brought by complex backgrounds. In some special cases, the regression loss adopted by the model cannot effectively guide the regression of the model, so the detection accuracy of the model is low.

3. Improvements to the YOLOX Model

This section analyses the shortcomings of the regression loss used in the original model and proposes a new regression loss function, SIoU-d. Meanwhile, the structure of the model is modified by replacing SPP with SPPF which slightly reduces the computational effort. Furthermore, PAN (path aggregation network) is replaced by FPN (feature pyramid) and ECA is embedded between the backbone and feature fusion layers to lighten the effect of the complex background. Figure 2 shows the structure of the improved model.

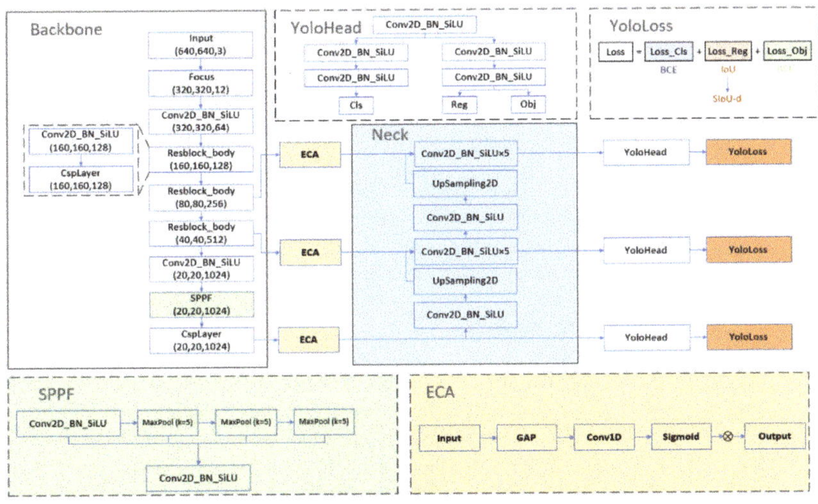

Figure 2. The structure of the improved model.

3.1. IOU Loss Analysis and Improvement

Insulator defect targets are small targets with low proportion and small scale in the input image. In the actual detection process, a small amount of offset and scaling of the predicted box can make significant impact on the detection accuracy of the model for small targets. Therefore, a suitable regression loss function would be important because it could effectively optimize the regression performance of the model and improve the detection accuracy of the model for defective small targets. At present, IoU (Intersection over Union) is often used as the evaluation index of the model edge regression effect. According to it, Yu et al. [25] proposed IoU loss. Moreover, regression losses such as GIoU [26], DIoU [27], CIoU [27], and EIoU [28] were proposed in the subsequent development.

The YOLOX-S model adopts IoU Loss as the regression loss of the model, and this loss is consistent with the evaluation index of the border regression which can guide the direction of model optimization to some extent. However, there are some problems. First of all, as is shown in Figure 3a, the two boxes have no intersection, and the value of IoU Loss is always 1, which means that it cannot effectively guide the optimization direction of the model. Secondly, when the same predicted box is in different positions within the ground-truth box, the value of IoU Loss remains the same, which cannot play a positive role in the optimization of the model, as in Figure 3b. In addition, predicted boxes of different shapes may have the same loss value within the same ground-truth box, as in Figure 3c. Finally, IoU does not specify the regression angle of the model, and the high degree of freedom in regression hinders the fast and accurate convergence of the model.

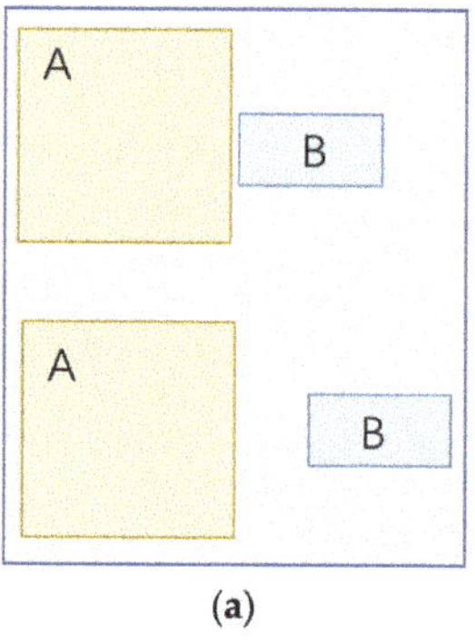

(a)

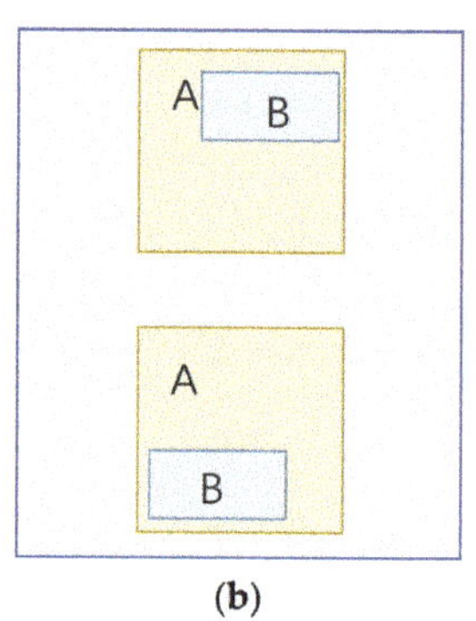

(b)

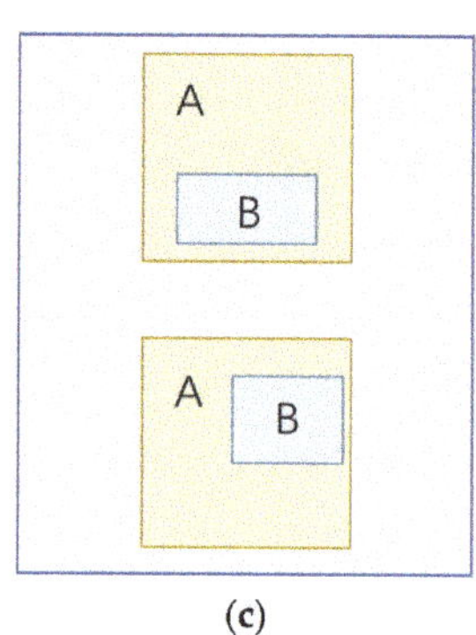

(c)

Figure 3. The shortcoming of IoU, A refers to the ground-truth box, and B refers to the predicted box: (**a**) the predicted box does not intersect with the ground-truth box; (**b**) the same predicted box is in different positions within the ground-truth box; (**c**) different predicted boxes with the same area in the ground-truth box.

As mentioned above, Gevorgyan [29] proposed the SIoU loss function. The new penalty terms are introduced based on IoU. Firstly, the x-axis component and y-axis component of the centroid distance are compared with the width and height of the smallest external rectangle. Then, the scale-insensitive information of the centroid distance on the x-axis and y-axis would be obtained which speeds up the regression of the model and improves the regression accuracy of the model. Secondly, the angles formed by the centroids of the two boxes and the x and y axes are calculated, and the angle loss is used to guide the centroids of the predicted boxes to regress along the x and y axes of the ground-truth box centroids. It reduces the freedom of the regression and further accelerates the convergence of the network. Finally, the width and height of the two boxes were compared separately. The scale-insensitive information of width and height was obtained, and it could further improve the regression accuracy of the model.

The SIoU performs in the optimization process of the predicted box with the center point of the predicted box converging to the x-axis. To be specific, it adjusts the center point of the predicted box approximately to the *x*-axis with the center point of the ground-truth box and further reduces the distance components of the two center points in the *x*-axis direction when the shapes of the two boxes are similar. In this process, the width and height of the predicted box and the distance components of the two center points in the *y*-axis direction are continuously adjusted to make the predicted box coincide with the actual box as much as possible.

Figure 4 shows the different convergence of the predicted box to the ground-truth box along the *x*-axis direction and the diagonal direction of the ground-truth box, respectively. In both figures, the width of the predicted box and the ground-truth box are *a*, the height is *b*, the original distance between the center points of both boxes is *d*, and the convergence rate is *c*. Figure 4a shows the convergence of the predicted box to the ground-truth box along the *x*-axis direction. The predicted box converges to the ground-truth box in two consecutive times, and the increment of the degree of overlap between the two boxes is $b \times c$. Figure 4b shows the convergence of the predicted box to the ground-truth box along the diagonal direction of the ground-truth box, and the increment of the degree of overlap of the two boxes is growing in the course of the two successive converged actual boxes. The increment of the degree of overlap is $\frac{abc^2}{a^2+b^2}$ in the first convergence process and $\frac{(3abc^2)}{a^2+b^2}$ in the second convergence process. Therefore, forcing the predicted box to converge along the diagonal direction of the ground-truth box can effectively increase the regression efficiency of the model and enable the loss function to converge quickly in the later stages of model optimization. Meanwhile, the convergence along the diagonal direction of the ground-truth box can make the center point of the predicted box on the x, y axes fall simultaneously in a certain proportion, which is more conducive to the optimization of the model.

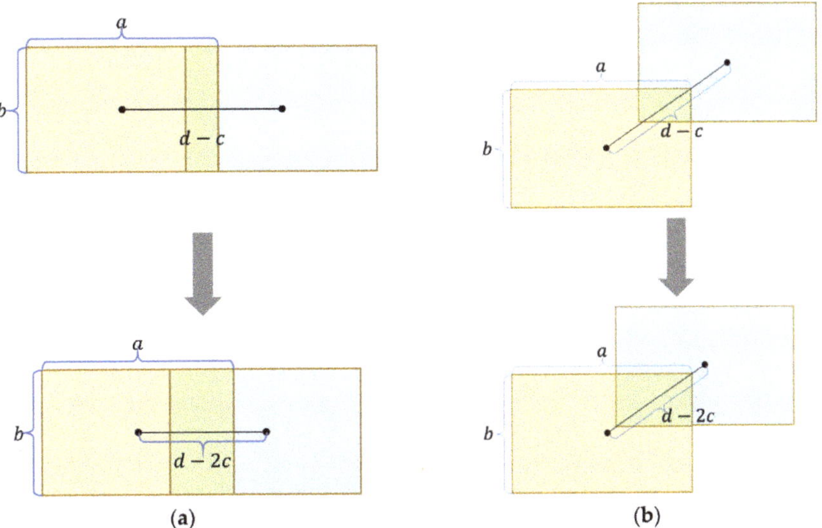

Figure 4. The yellow box refers to the ground-truth box, the blue box refers to the predicted box, the green box refers to the area where two boxes coincide. (**a**) The predicted box converges along the *x*-axis direction to the ground-truth box; (**b**) the predicted box converges along the diagonal direction of the ground-truth box to the ground-truth box.

In this paper, the angle loss of SIoU is improved, and SIoU-d is defined as follows:

$$L_{SIoU-d} = 1 - IoU + \frac{\Delta + \Omega}{2} \tag{1}$$

$$\Delta = 2 - e^{(\Lambda - 2) \times (\frac{c_w}{C_w})^2} - e^{(\Lambda - 2) \times (\frac{c_h}{C_h})^2} \tag{2}$$

$$\Omega = \left(1 - e^{-\frac{|w-w^{gt}|}{\max(w,w^{gt})}}\right)^{\theta} + \left(1 - e^{-\frac{|h-h^{gt}|}{\max(h,h^{gt})}}\right)^{\theta} \tag{3}$$

$$\Lambda = \cos(\beta - \alpha) \tag{4}$$

where Δ and Ω are the distance loss and shape loss. As is shown in Figure 5, the yellow box is the ground-truth box, and the blue box is the predicted box. c_w and c_h are the width and height of the rectangle constructed at the center of the two boxes, C_w and C_h are the width and height of the minimum external rectangle, w and h are the width and height of the predicted box, and w_{gt} and h_{gt} are the width and height of the ground-truth box. α refers to the angle between the center point of the predicted box and the center point of the ground-truth box, and β refers to the diagonal angle of the ground-truth box.

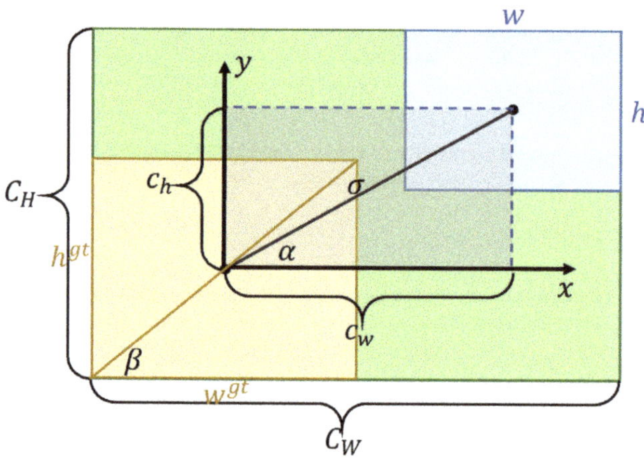

Figure 5. Schematic diagram of SIoU. The yellow box refers to the ground-truth box, the blue box refers to the predicted box, and the green area refers to the minimum external rectangle of the two boxes.

3.2. Analysis of Feature Fusion and Improvement of Embedded Attention Mechanism

The backbone of the model could extract a large amount of feature information. As the depth of the model increases, the model could extract not only the shallow location features but also the deep semantic features. In the meantime, it is necessary to combine the shallow location information with the deep semantic information to enhance the detection of the model at different scales.

The YOLOX model adopts PANet as the feature fusion layer. Firstly, the semantic features of the deep layer are passed to the shallow layer by up-sampling. Then the fused feature information is passed to the deep layer by down-sampling which increases both the semantic expression capability of the shallow layer and the localization capability of the deep layer. However, this feature fusion approach might be complicated for the detection of insulator defects, so this paper uses FPN as the feature fusion layer of the model and outputs the feature information from the deep layer directly to the prediction network of the model.

However, the information extracted by the backbone contains both valid feature information and invalid redundant features. The process of feature fusion cannot effectively reduce the impact of redundant information on the detection capability of the model, while the attention mechanism can assign greater weight to important features, reduce the weight of redundant information, and lighten the influence of redundant features [30]. Therefore, it is necessary to use the attention mechanism to process the extracted feature information and assign different weights to the feature information before the model undergoes feature fusion.

The attention mechanism originates from the study of human vision, which can selectively focus on important information. Channel attentions such as ECA (Efficient Channel Attention Module) and SE (Squeeze and Excitation) could assign weights to the features of each channel, it can improve the classification ability of the model to a certain extent. Spatial attention such as Non-Local Block could assign weights to the region where the target is located which can reduce the influence of the background and enhance the regression ability of the model. Moreover, there is a mixture of the above two types of attention, CBAM (Convolutional Block Attention Module). To reduce the impact of redundant information and keep the high detection speed of the model, ECA was finally selected to enhance the weights of important features before feature fusion.

The structure of the ECA module is shown in Figure 6, and its generated channel weights can correspond to the channels of the input feature information which can effectively improve the learning efficiency of the model. The value of k in the figure is adaptively related to the number of channels and is defined as follows:

$$k = \Psi(C) = \left| \frac{\log_2(C) + b}{\gamma} \right|_{odd} \tag{5}$$

where C represents the number of channels of the input feature information, and k is the nearest odd number of C which has been processed, $b = 1$, $\gamma = 2$.

ECA

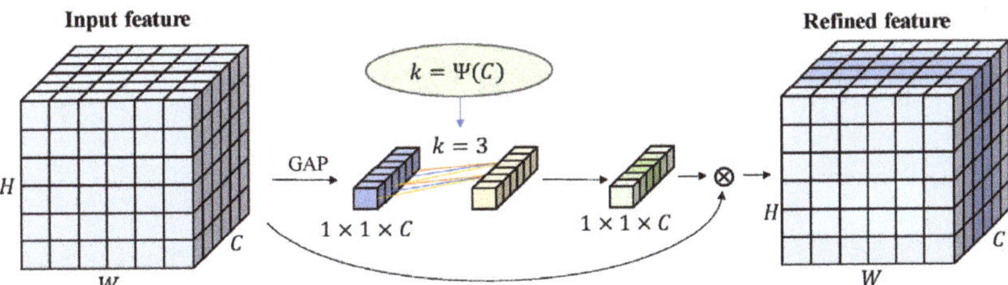

Figure 6. The module structure of ECA.

4. The Experiment and Evaluation Indexes

4.1. Experimental Conditions

Experiments were performed on a local Ubuntu 20.04.2 computer with 96 GB memory, CPU (Intel Xeon Platinum 8171M@ 2.60 GHz), 2 GPUs (NVIDIA GeForce RTX3090, 24 GB), and environment version is Pytorch 1.6.0.

This experiment used 1588 aerial images of insulators, including 647 images with defects. There were 2908 insulators in total and 715 insulators with defects. In the experiment, the images of the entire data set were scrambled, and the data set was divided into the training set, validation set, and test set according to the ratio of 8:1:1. Finally, 1286 images of the training set, 143 images of the validation set, and 159 images of the test set were obtained. The number of insulators and defects in each set is shown in Table 1.

Table 1. Data set division.

Dataset	Image	Insulator	Defect
Train	1286	2346	578
Val	143	257	65
Test	159	305	72
Total	1588	2908	715

The aerial images which have the defect of insulators are shown in Figure 7.

Figure 7. Images (**a**–**f**) show the defect of insulators.

4.2. Evaluation Indexes

In this experiment, three indicators of *mAP* (mean Average Precision), FPS (Frames Per Second) are used to evaluate the model. *AP* refers to the area of the curve enclosed by the prediction accuracy and recall of the model for a certain type of target. the definitions of *AP* and *mAP* are

$$AP = \int_0^1 P(\mathrm{R})\mathrm{dR} \tag{6}$$

$$mAP = \frac{\sum_{i=1}^k AP_i}{k} \tag{7}$$

P(R) refers to the Precision–Recall Curve, *k* represents the number of classes. Equations (8) and (9) show the calculation of precision and recall.

$$Precision = \frac{TP}{TP + FP} \tag{8}$$

$$Recal = \frac{TP}{TP + FN} \tag{9}$$

TP (True Positive) is the example that represents the positive sample that is correctly classified. *FP* (False Positive) means the negative sample that was misclassified. *FN* (False Negative) is the example that represents a misclassified positive sample.

4.3. Experimental Process

The experiments adopt the idea of transfer learning to train the model with pre-trained weights for 300 epochs. The first 50 epochs freeze the backbone of the model with the initial learning rate set to 1×10^{-3} and the batch size set as 16. After 50 epochs, the whole model is trained with the initial learning rate reduced to 1×10^{-4} and the batch size set as 8. To facilitate the comparison of the effects of different improvements on the experimental results, only one variable was changed for each training. The variables of the study included the regression angle of regression loss and the attention mechanism (SE, ECA, CBAM, and Non-Local).

Figure 8 shows the loss curves of the original YOLOX-S model and the improved model in this paper. In the figure, the validation loss of the original model converged to 2.6, and the validation loss of the improved model converged to 1.9. The trends gradually became stable in the subsequent training processes without overfitting.

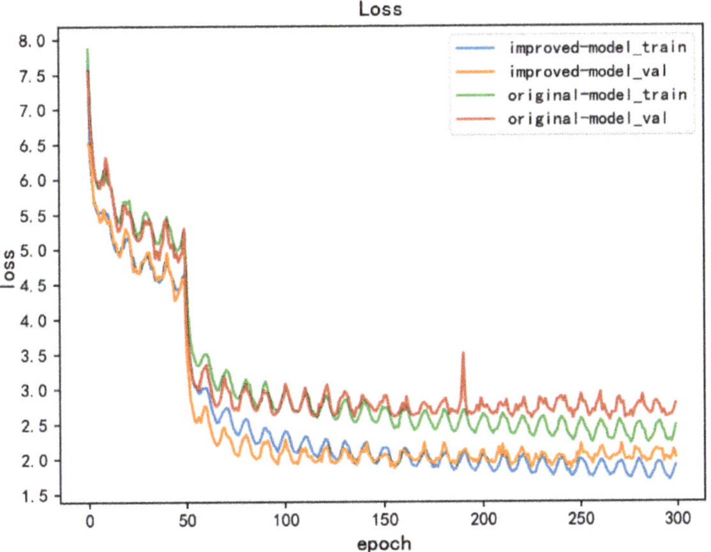

Figure 8. Loss curves of the original model and the improved method.

5. Research on Model Optimization Methods

5.1. The Effect of the Angle of Regression Loss

The regression angle is defined as the angle between the center point of the predicted box and the center point of the ground-truth box. Figure 9 illustrates the validation loss of the model with different regression angles. Test1 defines the regression angle as the diagonal angle of the ground-truth box, so that the center point of the predicted box regresses along the diagonal direction of the ground-truth box. Test2 defines the regression angle as $\pi/4$, so that the center point of the predicted box regresses to the center point of the ground-truth box along the line where $\pi/4$ is located. Test3 defines the regression angle as 0, so that the predicted box is regressed along the x-axis and y-axis direction where the center point is located.

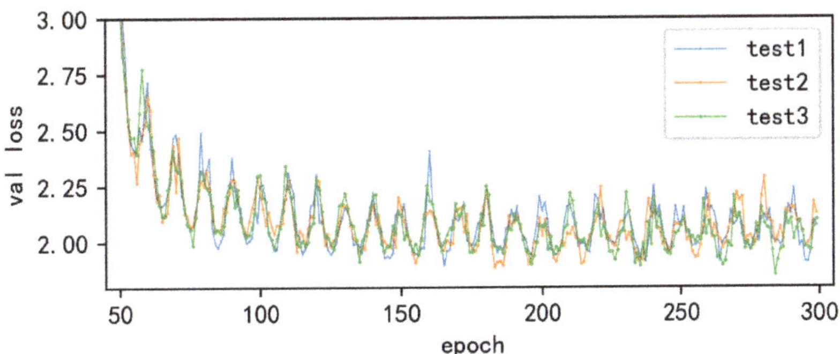

Figure 9. Validation loss of training with different regression angles.

During the training of test1, test2, and test3, the number of training epochs, in which the validation loss of training is less than 2, for the first time is 85, 114, and 116, respectively. It is concluded that the test1 has the fastest loss decrease and better training effect during the training.

Table 2 shows the detection accuracy of the models trained with these three regression angles with AP50. Test3 has the lowest detection AP for both targets. Test2 has the highest detection AP for defects, but it reduces the AP of the model to detect insulators. Test1 has a slightly lower detection AP than test2 for defects while it has the largest mAP, 94.51%. Overall, test1 performs the best with improved detection AP for both targets. Compared with the original model, the detection AP for defects is increased by 2.74%, and the detection AP for insulators is increased by 0.18%.

Table 2. Detection AP of the models trained with different regression angles at AP50.

Method	Defect AP/%	Insulator AP/%	mAP/%
Base [1]	92.76	96.12	94.44
Test1	95.39	96.30	95.84
Test2	95.44	96.04	95.74
Test3	95.33	95.96	95.64

[1] The base model is YOLOX-S.

The experimental results show that the closer the selected regression angle is to the angle of the diagonal of the ground-truth box, the faster the convergence of the model, the higher the detection accuracy. Therefore, the angle of the diagonal of the ground-truth box is selected as the angle of the model regression loss in this paper.

5.2. The Impacts of Attention Mechanism

To investigate the influence of attention mechanism on model detection accuracy, this paper embedded SE, ECA, CBAM, and Non-Local between the backbone and feature fusion layers of the model and trained the model separately.

Table 3 shows the detection AP of the models with these four attention mechanisms at AP50. The model detection mAP with embedded SE, ECA, CBAM, and Non-Local is 94.74%, 95.17%, 95.10%, and 94.40% where ECA and CBAM performed similarly. CBAM is slightly better than ECA in defect detection. It is mainly due to the reason that CBAM has both channel and spatial attention. Spatial attention has an enhanced effect on the regression effect of the predicted box, but this attention mechanism has a more complex computational process compared to ECA and has a greater impact on the detection speed of the model. Considering these factors, ECA has a better performance with the AP increased to 0.59% for insulators and 0.86% for defect AP. It can effectively assign weights to the feature information and improve the detection effect of the model.

Table 3. Detection AP of the models with different attention mechanisms at AP50.

Method	Defect AP/%	Insulator AP/%	mAP/%	Fps
SE	93.52	95.92	94.74	71
ECA	93.62	96.71	95.17	71
CBAM	93.89	96.31	95.10	66
Non-Local	93.34	95.46	94.40	33

Figure 10 shows the heat map of the original model, and Figure 11 shows the heat map of the model with ECA embedded. Among them, the center of the target which is focused by the model would be highlighted. The higher the brightness, the more attention it receives. The green box represents the insulator detected, and the orange box represents the insulator defect detected. Only some of the highlight insulators in Figure 10a,b are detected. In addition, the insulator defects in Figure 10c receive almost no attention from the model. Figure 11a,b can detect the low confidence insulators in Figure 10a,b which received attention from the model but were not detected. As for the insulator defects that existed in Figure 10c, the model pays high attention and detected them successfully. The results indicate that the embedding of the attention mechanism can effectively mitigate the influence of redundant features, improve the sensitivity of the model to important features of the target, and improve performance.

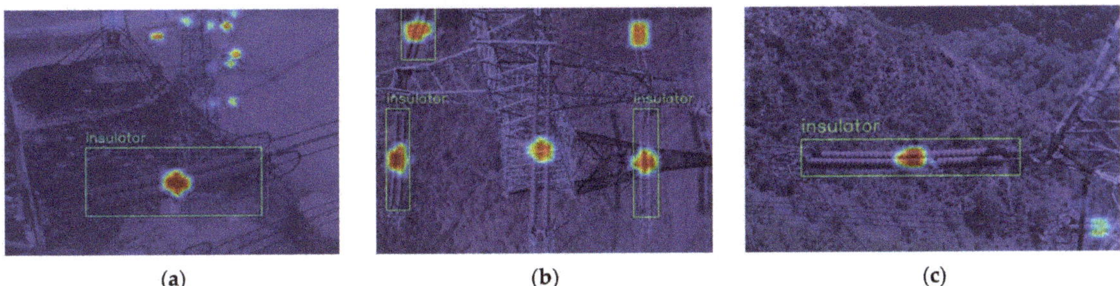

Figure 10. Heat map of the detection of the original YOLOX-S model: (**a**) only one insulator has been detected; (**b**) three insulators have been detected; (**c**) only one insulator has been detected, and the defect has not been detected.

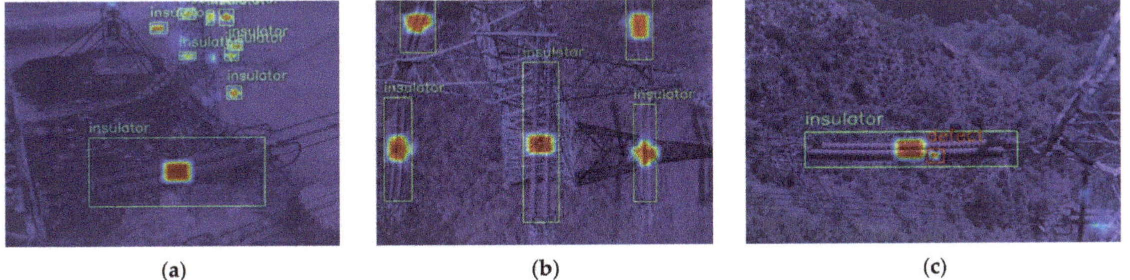

Figure 11. Heat map of the detection with the ECA embedded: (**a**) nine insulators have been detected; (**b**) five insulators have been detected; (**c**) one insulator and one defect have been detected.

5.3. The Comparison of Predicted Results

As is shown in Table 4, ablation experiments are used to verify the effectiveness of the improved algorithm in this paper. The position of "$\sqrt{}$" in the table indicates that the algorithm adopts this improved strategy. Algorithm 3 introduces CBAM on the basis of Algorithm 2. This attention mechanism has little effect on the improvement of model detection AP. The detection AP of insulators is increased by 0.1%, and the defect detection AP is increased by 0.51%.

Table 4. The ablation experiments of different improvement strategy.

Model	FPN	SIoU-d	ECA	CBAM	Insulator AP/%	Defect AP/%	mAP/%	Fps
YOLOX-S					96.12	92.76	94.44	74
Algorithm 1		$\sqrt{}$			96.30	95.39	95.84	72
Algorithm 2	$\sqrt{}$	$\sqrt{}$			96.28	96.28	96.28	71
Algorithm 3	$\sqrt{}$	$\sqrt{}$		$\sqrt{}$	96.38	96.79	96.58	62
Algorithm 4	$\sqrt{}$	$\sqrt{}$	$\sqrt{}$		96.57	97.79	97.18	71

In comparison, ECA is introduced in Algorithm 4 on the basis of Algorithm 2. The introduction of this attention mechanism greatly improves the detection AP of the model for small targets with defects. The insulator detection AP is increased by 0.29%, and the defect detection AP is increased by 1.51%. This result shows that ECA has a better performance in improving the detection effect of the improved model in this paper.

Based on the above studies, this paper proposed an improved YOLOX-S model by using SIoU-d as the regression loss of the model to enhance the regression performance of the model and ECA to lighten the influences of redundant features on the detection accuracy of the model.

Figure 12 shows the predicted results of the original model, and Figure 13 shows the predicted results of the improved model, where the red boxes indicate the detected insulator targets, and the blue boxes indicate the detected defect targets. Only three insulators were detected in Figure 12a. It is clear that insulators with complex backgrounds and insulators with similar colors to the background cannot be detected well. One insulator was detected in Figure 12b, and the defects present on it and the blocked insulator were not detected. Two insulators were detected in Figure 12c, but the insulator which is located below it was not completely boxed out. Figure 13a detects the two insulators that could not be detected inFigure 12a.; Figure 13b detects the defect present on the insulator and the insulator that is blocked in the lower right corner. Figure 13c detects the defect located on the insulator below while completely boxing out the insulator below.

(a)   (b)   (c)

Figure 12. Detection result of the original model: (**a**) three insulators have been detected; (**b**) only one insulator has been detected, and the defect has not been detected; (**c**) two insulators have been detected, and the defect has not been detected.

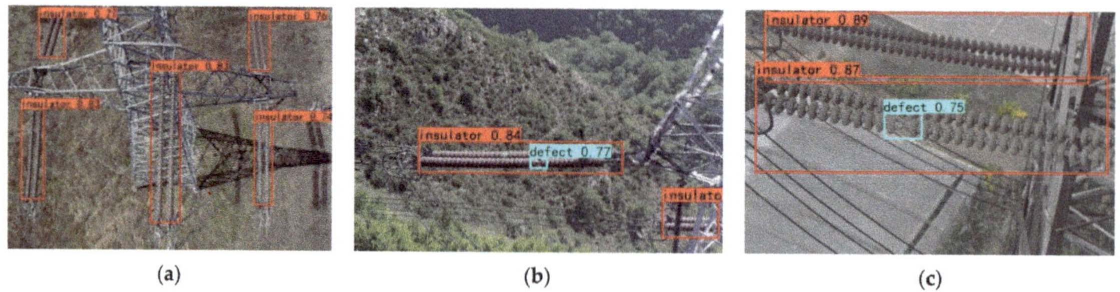

(a)   (b)   (c)

Figure 13. The detection result of the improved model: (**a**) five insulators have been detected; (**b**) two insulators and one defect have been detected; (**c**) two insulators and one defect have been detected.

The images adopted in the prediction of experiments are the actual detection images which were not used in the training. The results show that the improved model could effectively lighten the influence of the complex background on the detection of insulators and its defects. The model could accurately detect insulator defects and have the ability of generalization.

Table 5 shows the comparison of the improved model with other models at AP50. The Algorithm 4 achieves the best results in the detection of insulators and defective targets, with 96.57% AP for insulator detection which increased by 0.45% and 97.79% AP for defective targets which increased by 5.03%. In addition, there is no obvious decrease in its detection speed.

Table 5. Detection accuracy of different models at AP50.

Model	Insulator AP/%	Defect AP/%	mAP/%	Fps
Faster-RCNN	93.24	65.05	79.15	8
SSD	86.74	62.04	74.39	65
YOLOv3	93.62	89.68	91.65	39
YOLOv4	91.86	90.43	91.15	32
YOLOv5-S	92.54	93.03	92.78	71
YOLOX-S	96.18	92.93	94.55	74
Algorithm 4	96.57	97.79	97.18	71

6. Conclusions

To achieve intelligent inspection of transmission lines, lighten the influence of complex background on model detection, and improve the detection effect of the model on defects, this paper explores the effects of regression angle and attention mechanism on model accuracy based on YOLOX-S. The experimental results show that the regression of the predicted box along the diagonal direction of the ground-truth box can effectively enhance the regression effect of the model and improve the detection accuracy of the model for small targets. The embedded of the channel attention mechanism between the backbone and feature fusion layers can effectively lighten the influence of the complex background on the detection accuracy of the model. The improved model in this paper has a detection AP of 97.79% for insulator defects, achieving a rise of 5.03%. In addition, detection AP reached 96.57% for insulators, a rise of 0.45%. The detection speed rose to 71 fps which can satisfy the purpose of fast and accurate detection of defective small targets.

Author Contributions: Conceptualization, G.H. and T.L.; methodology, G.H.; software, T.L.; validation, T.L., M.Z. and R.W.; formal analysis, G.H.; investigation, Q.L. and F.Z.; resources, K.L., Q.L. and F.Z.; data curation, Q.Y., Q.L. and F.Z.; writing—original draft preparation, T.L.; writing—review and editing, G.H., T.L. and L.Q.; visualization, M.Z.; supervision, K.L. and L.Q. All authors have read and agreed to the published version of the manuscript.

Funding: This work was supported by the National Key R & D Program of China (No. 2020YFB0905900).

Institutional Review Board Statement: Not applicable.

Informed Consent Statement: Not applicable.

Data Availability Statement: Dataset link: https://github.com/InsulatorData/InsulatorDataSet (accessed on 18 May 2022).

Conflicts of Interest: The authors declare no conflict of interest.

References

1. Qiu, Z.B.; Yu, X.B.; Huo, F.; Liu, Z.; Gong, W.X.; Li, Y.Q. Spray Image Processing of Composite Insulators Based on Interval Classification of Uniformity Measure and Intelligent Identification of Hydrophobicity. *High Volt. Eng.* **2020**, *46*, 3008–3017.
2. Tan, P.; Li, X.F.; Xu, J.M.; Ma, J.E.; Wang, F.J.; Ding, J.; Fang, Y.T.; Ning, Y. Catenary insulator defect detection based on contour features and gray similarity matching. *J. Zhejiang Univ.-SCIENCE A* **2020**, *21*, 64–73. [CrossRef]
3. Yin, J.G.; Lu, Y.P.; Gong, Z.X.; Jiang, Y.C.; Yao, J.G. Edge Detection of High-Voltage Porcelain Insulators in Infrared Image Using Dual Parity Morphological Gradients. *IEEE Access* **2019**, *7*, 32728–32734. [CrossRef]
4. Tang, B.; Qin, Q.; Huang, L. Transmission line aerial image recognition of insulator strings based on color model and texture features. *J. Electr. Power Sci. Technol.* **2020**, *35*, 13–19.

5. Tang, W.H.; Niu, Z.W.; Zhao, B.N.; Ji, T.Y.; Li, M.S.; Wu, Q.H. Research and Application of Data-driven Artificial Intelligence Technology for Condition Analysis of Power Equipment. *High Volt. Eng.* **2020**, *46*, 2985–2999.
6. Zhang, Y.H.; Qiu, C.M.; Yang, F.; Xu, S.W.; Shi, X.; He, X. Overview of Application of Deep Learning with Image Data and Spatio-temporal Data of Power Grid. *Power Syst. Technol.* **2019**, *43*, 1865–1873.
7. Tao, X.; Zhang, D.; Wang, Z.; Liu, X.; Zhang, H.; Xu, D. Detection of power line insulator defects using aerial images analyzed with convolutional neural networks. *IEEE Trans. Syst. Man Cybern. Syst.* **2018**, *50*, 1486–1498. [CrossRef]
8. Ling, Z.N.; Zhang, D.X.; Qiu, R.C.; Jin, Z.J.; Zhang, Y.H.; He, X.; Liu, H.C. An accurate and real-time method of self-blast glass insulator location based on faster R-CNN and U-net with aerial images. *CSEE J. Power Energy Syst.* **2019**, *5*, 474–482. [CrossRef]
9. Li, X.F.; Su, H.S.; Liu, G.H. Insulator Defect Recognition Based on Global Detection and Local Segmentation. *IEEE Access* **2020**, *8*, 59934–59946. [CrossRef]
10. Liang, H.G.; Zuo, C.; Wei, W.M. Detection and Evaluation Method of Transmission Line Defects Based on Deep Learning. *IEEE Access* **2020**, *8*, 38448–38458. [CrossRef]
11. Zhai, Y.J.; Yang, X.; Wang, Q.M.; Zhao, Z.B.; Zhao, W.Q. Hybrid Knowledge R-CNN for Transmission Line Multifitting Detection. *IEEE Trans. Instrum. Meas.* **2021**, *70*, 1–12. [CrossRef]
12. Zhao, W.Q.; Xu, M.F.; Cheng, X.F.; Zhao, Z.B. An Insulator in Transmission Lines Recognition and Fault Detection Model Based on Improved Faster RCNN. *IEEE Trans. Instrum. Meas.* **2021**, *70*, 1–8. [CrossRef]
13. Wang, S.Q.; Liu, Y.F.; Qing, Y.H.; Wang, C.X.; Lan, T.Z.; Yao, R.T. Detection of Insulator Defects with Improved ResNeSt and Region Proposal Network. *IEEE Access* **2020**, *8*, 184841–184850. [CrossRef]
14. Wang, B.; Dong, M.; Ren, M.; Wu, Z.Y.; Guo, C.X.; Zhuang, T.X.; Pischler, O.; Xie, J.C. Automatic Fault Diagnosis of Infrared Insulator Images Based on Image Instance Segmentation and Temperature Analysis. *IEEE Trans. Instrum. Meas.* **2020**, *69*, 5345–5355. [CrossRef]
15. Zhong, J.P.; Liu, Z.G.; Cheng, Y.; Wang, H.R.; Gao, S.B.; Núñez, A. Adversarial Reconstruction Based on Tighter Oriented Localization for Catenary Insulator Defect Detection in High-Speed Railways. *IEEE Trans. Intell. Transp. Syst.* **2020**, *23*, 1109–1120. [CrossRef]
16. Sadykova, D.; Pernebayeva, D.; Bagheri, M.; James, A. IN-YOLO: Real-Time Detection of Outdoor High Voltage Insulators Using UAV Imaging. *IEEE Trans. Power Deliv.* **2019**, *35*, 1599–1601. [CrossRef]
17. Wang, Q.; Yi, B.S. Insulator Defect Recognition in Aerial Images Based on Gaussian YOLOv3. *Laser Optoelectron. Prog.* **2021**, *58*, 254–260.
18. Zhang, X.T.; Zhang, Y.Y.; Liu, J.F.; Zhang, C.H.; Xue, X.Y.; Zhang, H.; Zhang, W. InsuDet: A Fault Detection Method for Insulators of Overhead Transmission Lines Using Convolutional Neural Networks. *IEEE Trans. Instrum. Meas.* **2021**, *70*, 1–12. [CrossRef]
19. Shen, J.Q.; Chen, X.J.; Zhai, H.X. YOLOv3 Detection Algorithm Based on the Improved Bounding Box Regression Loss. *Comput. Eng.* **2022**, *48*, 236–243. [CrossRef]
20. Duan, C.H.; Wang, X.F.; Ji, L.J.; Cao, R.N. Application research of steel coil end face defect detection based on improved YOLOv3. *Manuf. Autom.* **2021**, *43*, 185–188.
21. Tang, X.Y.; Huang, J.B.; Feng, J.W.; Chen, X.H. Image Segmentation and Defect Detection of Insulators Based on U-net and YOLOv4. *J. South China Norm. Univ. (Nat. Sci. Ed.)* **2020**, *52*, 15–21.
22. Lv, F.C.; Niu, L.L.; Wang, S.H.; Xie, Q.; Wang, Z.H. Intelligent Detection and Parameter Adjustment Strategy of Major Electrical Equipment Based on Optimized YOLOv4. *Trans. China Electrotech. Soc.* **2021**, *36*, 4837–4848. [CrossRef]
23. Qiu, Z.; Zhu, X.; Liao, C.; Shi, D.; Qu, W. Detection of Transmission Line Insulator Defects Based on an Improved Lightweight YOLOv4 Model. *Appl. Sci.* **2022**, *12*, 1207. [CrossRef]
24. Ge, Z.; Liu, S.T.; Wang, F.; Li, Z.M.; Sun, J. YOLOX: Exceeding YOLO Series in 2021. *arXiv* **2021**, arXiv:2107.08430.
25. Yu, J.H.; Jiang, Y.N.; Wang, Z.Y.; Cao, Z.M.; Huang, T. Unitbox: An advanced object detection network. In Proceedings of the 24th ACM International Conference on Multimedia, Amsterdam, The Netherlands, 15–19 October 2016; pp. 516–520. [CrossRef]
26. Rezatofighi, H.; Tsoi, N.; Gwak, J.Y.; Sadeghian, A.; Reid, L.; Savarese, S. Generalized intersection over union: A metric and a loss for bounding box regression. In Proceedings of the IEEE/CVF Conference on Computer Vision and Pattern Recognition, Long Beach, CA, USA, 15–20 June 2019; pp. 658–666.
27. Zheng, Z.H.; Wang, P.; Liu, W.; Li, J.Z.; Ye, R.G.; Ren, D.W. Distance-IoU loss: Faster and better learning for bounding box regression. In Proceedings of the AAAI Conference on Artificial Intelligence, New York, NY, USA, 7–12 February 2020; Volume 34, pp. 12993–13000. [CrossRef]
28. Zhang, Y.F.; Ren, W.Q.; Zhang, Z.; Jia, Z.; Wang, L.; Tan, T.N. Focal and efficient IOU loss for accurate bounding box regression. *arXiv* **2021**, arXiv:2101.08158. [CrossRef]
29. Gevorgyan, Z. SIoU Loss: More Powerful Learning for Bounding Box Regression. *arXiv* **2022**, arXiv:2205.12740.
30. Guo, M.H.; Xu, T.X.; Liu, J.J.; Liu, Z.N.; Jiang, P.T.; Mu, T.J.; Zhang, S.H.; Martin, R.R.; Cheng, M.M.; Hu, S.M. Attention Mechanisms in Computer Vision: A Survey. *Comput. Vis. Media* **2022**, *8*, 331–368. [CrossRef]

Article

Multi-Geometric Reasoning Network for Insulator Defect Detection of Electric Transmission Lines

Yongjie Zhai [1], Zhedong Hu [1], Qianming Wang [1], Qiang Yang [2,*] and Ke Yang [1]

1   Automation Department, North China Electric Power University, Baoding 071003, China
2   College of Electrical Engineering, Zhejiang University, Hangzhou 310027, China
*   Correspondence: qyang@zju.edu.cn; Tel.: +86-0571-87953296

Abstract: To address the challenges in the unmanned system-based intelligent inspection of electric transmission line insulators, this paper proposed a multi-geometric reasoning network (MGRN) to accurately detect insulator geometric defects based on aerial images with complex backgrounds and different scales. The spatial geometric reasoning sub-module (SGR) was developed to represent the spatial location relationship of defects. The appearance geometric reasoning sub-module (AGR) and the parallel feature transformation (PFT) sub-module were adopted to obtain the appearance geometric features from the real samples. These multi-geometric features can be fused with the original visual features to identify and locate the insulator defects. The proposed solution is assessed through experiments against the existing solutions and the numerical results indicate that it can significantly improve the detection accuracy of multiple insulator defects using the aerial images.

Keywords: multi-geometric reasoning; insulator defect detection; deep learning; graph convolutional neural network

Citation: Zhai, Y.; Hu, Z.; Wang, Q.; Yang Q.; Yang, K. Multi-Geometric Reasoning Network for Insulator Defect Detection of Electric Transmission Lines. *Sensors* **2022**, 22, 6102. https://doi.org/10.3390/s22166102

Academic Editor: Paweł Pławiak

Received: 21 July 2022
Accepted: 13 August 2022
Published: 15 August 2022

1. Introduction

The electric power transmission line insulators are widely used in power systems to fix and insulate electrical equipment [1]. Due to the long-term exposure to the natural environment, the insulators are susceptible to damage, and the chance of failure of damaged insulators will increase, which will directly threaten the stability and safety of transmission lines [2,3]. Therefore, timely detection of insulator defects is crucial. Common insulator defects include damaged [4] and missing [5]. The geometric defect of an object is defined as the change of its geometry caused by natural or human factors, as shown in Figure 1.

Figure 1. Aerial insulator images with geometric defects (red circle). (**a,b**): missing defects; (**c,d**): damaged defects.

A common approach for insulator defect detection under these two categories is to capture aerial images of insulators by unmanned aerial vehicles (UAVs), which are then analyzed and processed using computer vision techniques [6–8]. Several recent studies have demonstrated the effectiveness of machine learning techniques in insulator detection. For instance, the authors in [9] achieved the detection of damaged insulator strings by extracting the shape and texture information of the insulator and improving the watershed algorithm. The study in [10] assessed the damage level of insulators by extracting insulator features with wavelet transform and then analyzing the insulator condition through the support vector machine. The authors in [11] represented the local features by introducing multi-scale and multi-feature descriptors, and then proposed a coarse-to-fine matching strategy to achieve insulator detection based on the spatial order features in the local features. Although these studies have achieved certain considerable results, machine learning approaches for vision processing by designing feature extraction modules are difficult to achieve better adaptability and robustness.

In recent years, advances in artificial intelligence enable deep-learning-based techniques to be adopted for insulator defect detection. In [12], the authors proposed an insulator missing detection network with compact feature space based on a stochastic configuration network and feedback transfer learning mechanism, and achieved adaptive adjustment of the depth feature space; The work in [13] introduced a deep neural network with cascade structure, which converted the insulator defect problem into a two-level object detection problem and realized the localization and identification of insulator missing. The authors in [14] improved the YOLOv3 model by employing Spatial Pyramid Pooling network (SPP) and multiscale prediction network and carried out training on a large number of insulator missing samples, achieving insulator defect detection under different aerial photography backgrounds. These insulator defect detections mainly focus on one single defect, i.e., missing transmission line insulators, and have achieved satisfactory results. Such a defect occurs frequently in overhead transmission lines in practice, and hence a large number of sample images are available that can be used for deep learning models.

Unfortunately, the sample images are scarce in the actual detection environment for defects such as damage d, flashover [15], and dirty [16]. Moreover, the morphological features presented by these defects are more diverse and complex than the missing defect, making it difficult for deep learning models to perform better. The work in [17] proposed a multiscale residual neural network that achieved insulator damage detection in a single background through rich spatial correlation and channel correlation. In [18] the authors implemented a state assessment for the existence of ice, snow, and water on the insulator surface based on the YOLOv2 model through data expansion. The authors in [19] realized the detection of insulator damaged defects by improving Regional Proposal Network (RPN) and adding the improved ResNeSt [20] for feature extraction. In general, the existing research is mainly from the perspective of data augmentation of defect samples, by increasing the sample quantity and then completing the training for the deep learning model to realize the detection of scarcity defects. However, the application of deep learning technology based on a small number of samples to defect detection of transmission line insulators is still at the exploratory stage. Furthermore, much research effort has been made in the field of machine learning-based detection methods. The current object detection frameworks are mainly divided into one-stage detectors and two-stage detectors. The one-stage detectors are mainly based on SSD [21] and YOLO series [22], and the representative frameworks presented by the two-stage detectors include Fast R-CNN [23], Faster R-CNN [24], and Mask R-CNN [25]. However, when used in practical industrial scenarios, even high-performance object detectors have difficulty showing great performance. For example, detectors with excellent performance on public datasets often are difficult to work reliably in power systems [26] or transportation systems [27] with complex scenarios.

However, there still remains a set of technical challenges in deep learning for defect detection of insulator s. The three challenges are as follows: (1) Complex background. As a basic electrical insulation device, insulators are widely applied in fields, woods, and buildings. (2) Various types. The existing insulators mainly include glass insulators, ceramic insulators, and composite insulators, which make the defects under different insulators more diverse. (3) Different scales. Insulator defects show multi-scale in appearance. For example, the scale of missing is relatively large, while the scale of damage is small. In order to better accomplish the detection of geometric defects of insulators. We propose a Multi-Geometric Reasoning Networks (MGRN), which fully taps into the geometric information of defect samples and the spatial location information of defects to address the three challenges in insulator defect detection. The detection accuracy of insulator geometric defects is significantly improved. The main technical contributions of this work can be summarized as follows:

1. Aiming at the different challenges in the defect detection of insulators, we construct two different types of geometric features and propose a multi-geometric reasoning network model (MGRN) to integrate them. This model can effectively improve the detection accuracy of insulator defects on transmission lines, and the recognition effect is remarkable, especially for some hard-detection geometric defects.
2. The appearance geometric reasoning (AGR) module is used to extract artificial defect sample features. The parallel feature transformation (PFT) module can enable the feature to be used in the real defect samples and extract the appearance geometric feature of the real defect samples. The spatial geometric reasoning (SGR) module is used to extract spatial geometric features of real defect samples. Thus, the multi-geometric features can be better integrated into the deep learning model.
3. The model can achieve better performance on a small number of samples, as well as a better improvement in insulator damage defect detection. It also provides a new idea for multi-scale object detection with few samples.

The rest of the paper is organized as follows. Section 2 describes the appearance geometric reasoning model, and the parallel feature transformation, the proposed spatial geometric reasoning model. Section 3 assesses the performance of the solution and presents the numerical results. Finally, the conclusive remarks are given in Section 4.

2. Proposed MGRN Based Solution

2.1. System Overview

In this article, insulator geometry defect detection is regarded as an instance-level task. The proposed MGRN is built on the Faster R-CNN [24] as the detector framework. The architecture of the proposed approach includes three major components: (1) appearance geometric reasoning sub-module (AGR) for artificial sample defect feature information extraction; (2) parallel feature transformation sub-module (PFT) for mapping artificial sample features to real sample feature space; (3) spatial geometric reasoning sub-module (SGR) for real sample space information extraction, as illustrated in Figure 2.

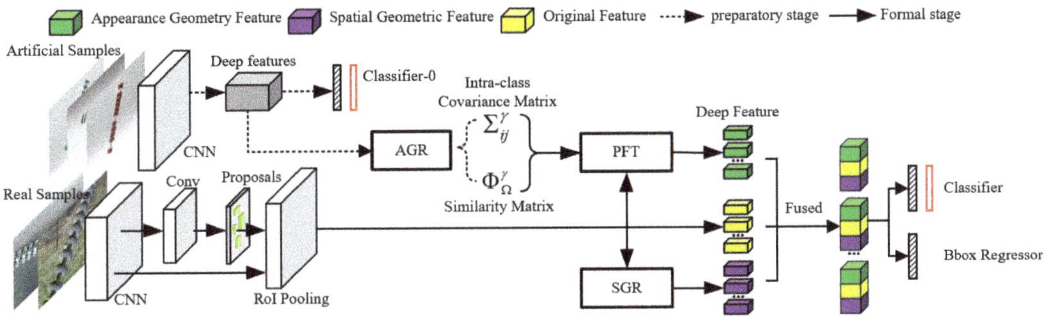

Figure 2. Architecture of the proposed MGRN. In the preparatory phase, the appearance geometric features of the artificial samples are modelled as intra-class covariance matrices and similarity matrices by the AGR sub-module. In the formal phase, the original features are extracted from the real samples (using CNN and RoI Pooling); the appearance geometric and spatial geometric features are extracted from the real samples (using SGR sub-module and PFT sub-module); finally the obtained features are fused from the real samples.

The MGRN network involves a preparatory phase and a formal phase. In the preparatory phase, the appearance geometric features of the artificial samples are modelled as intra-class covariance matrices and similarity matrices by the AGR sub-module. In the formal phase, our algorithm produces feature map and region proposals from input real images based on the CNN backbone and region proposal network (RPN). Next, the PFT and SGR sub-modules are utilized to respectively generate appearance geometric features and spatial geometric features. Finally, these feature maps are fused. The new method MGRN effectively introduces the geometric information from the space and exterior points of view and improves the accuracy of insulator defects detection.

2.2. Appearance Geometric Reasoning

In aerial images of insulator defects, due to the scarcity of defect sample data, it is difficult for data-driven deep learning models to achieve good results. Therefore, an appearance geometric reasoning sub-module is proposed which uses artificial defect samples to assist in extracting defect appearance features, as shown in Figure 3. A large number of artificial 2D solid-color background images are generated by constructing 3D artificial samples, as shown in Figure 4. These artificial defect images have the advantages of a simple background, and obvious and diverse defect features so that we can better obtain the appearance and geometric features of the defects.

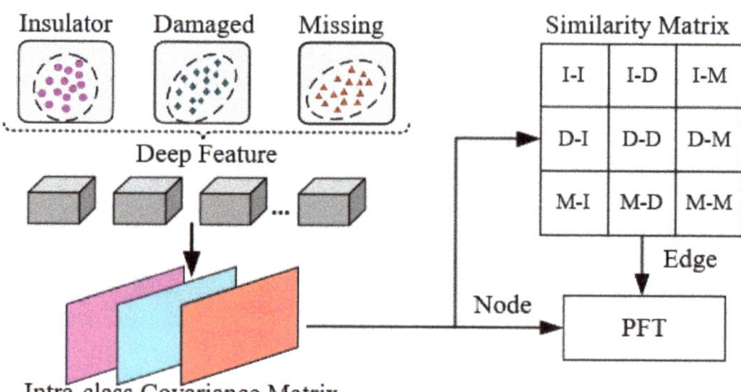

Figure 3. Architecture of the AGR model.

Figure 4. Artificial defect sample images of insulators (glass, ceramic and composite insulators).

First, the artificial defect images are delivered to a convolutional neural network for extracting deep features $f = F(x, \Theta_F) \in \mathbb{R}^{3 \times 512 \times 512}$, where x is the input image; $F(\bullet)$ is the parameterized feature extractor of Θ_F. The deep features f include the distribution of three main types of features, which are the main body features of the insulator (I), insulator damaged features (D), and insulator missing features (M). It is obvious that the feature similarity of I-D is greater than that of I-M and the feature similarity of I-M is greater than that of D-M in terms of physical geometry ($S_{I-D} > S_{I-M} > S_{D-M}$), as shown in Figure 5. It can be observed that whether the appearance of geometric features of artificial defect samples can be fully extracted depends on the training results of the feature extraction network. For this purpose, we choose the deep features with accuracy γ ($95\% \leq \gamma \leq 99\%$), and γ is a hyperparameter. Next, the intra-class covariance matrices Σ_{ij}^{γ} of the three types of features are calculated to represent the distribution of the appearance geometric features of the insulator body, insulator damaged, and insulator missing. The calculation formula is as follows:

$$\Sigma_{ij}^{\gamma} = cov(f_i^{\gamma}, f_j^{\gamma}), (i \neq j; i, j \in 1, 2, 3) \tag{1}$$

where $\Sigma_{ij}^{\gamma} = \{\Sigma_I^{\gamma}, \Sigma_D^{\gamma}, \Sigma_M^{\gamma} \big| \Sigma_{ij}^{\gamma} \in \mathbb{R}^{1 \times 512 \times 512}\}$.

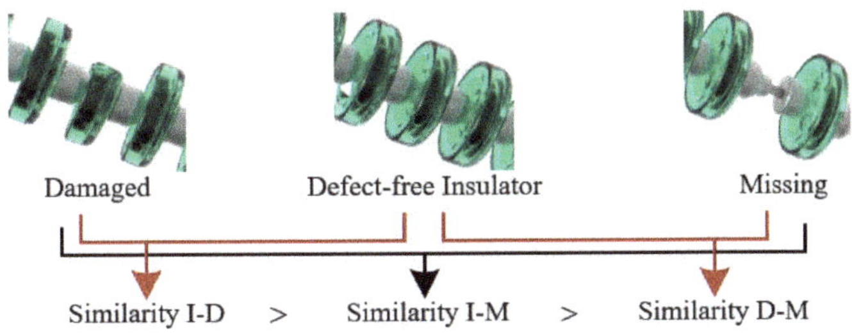

Figure 5. Similarity ranking of insulators with damaged and missing defects compared with the defect-free insulator.

Then, the cosine similarity $\Omega_{\alpha,\beta}^{\gamma}$ of the appearance geometric features Σ_I^{γ}, Σ_D^{γ} and Σ_M^{γ} are calculated to represent the geometric similarity of the physical appearance of the insulator and geometric defects on the instance-level image. The calculation formula is as follows:

$$\Omega_{\alpha,\beta}^{\gamma} = \frac{\Sigma_{\alpha}^{\gamma} \cdot \Sigma_{\beta}^{\gamma}}{\left\|\Sigma_{\alpha}^{\gamma}\right\|\left\|\Sigma_{\beta}^{\gamma}\right\|}, (\alpha, \beta \in \{I, D, M\}) \tag{2}$$

At this point, we get the two parameters delivered to the PFT sub-module of the formal phase for graph reasoning, which are node features Σ_I^{γ}, Σ_D^{γ} and Σ_M^{γ} (the appearance geometric features of artificially defect samples) and node relationship matrix Φ_{Ω}^{γ} (the similarity matrix of appearance geometric features). Finally, we ensure the convergence of the network parameters during the training of artificially defect samples by a cross-entropy loss function. The calculation formula is as follows:

$$L_1 = -\sum_{i=1}^{n} p(x_i) log(q(x_i)) \tag{3}$$

where n is the number of samples; $p(x_i)$ is the artificial defect sample label, and $q(x_i)$ is the probability of the defect category predicted by the model.

In this section of appearance geometric reasoning, we describe how to obtain information about the appearance geometric features and similarity of defect samples. However, the information is obtained based on artificial samples. How to apply the information to the detection of real defect samples is a problem worth considering. To solve the problem, the parallel feature transformation sub-module (PFT) is designed for Artificial-Real transformation.

2.3. Parallel Feature Transformation

In this section, the parallel feature transformation sub-module (PFT) will be described in detail, as shown in Figure 6.

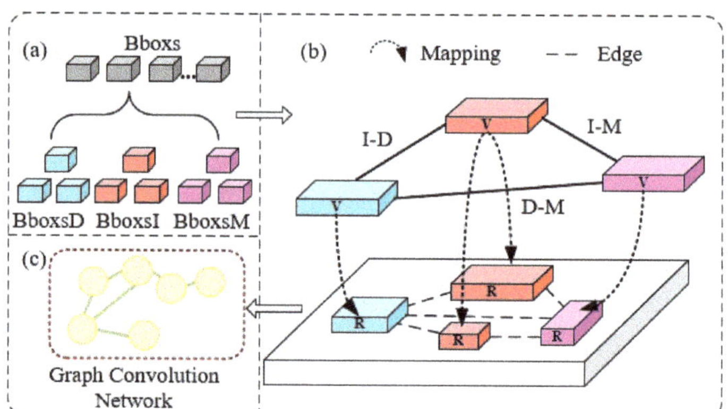

Figure 6. Architecture of the PFT model. (**a**) shows that the box contains category information and feature information; (**b**) shows a mapping mechanism; (**c**) shows the process of using graph neural network for the appearance geometric reasoning of real samples

In the formal phase, the object regional proposals (Bboxs) are obtained from the real samples through CNN, RPN, and RoI Pooling in turn. The Bboxs contain the category information and feature information, as shown in Figure 6a. Then, we map the appearance geometric features Σ_I^γ, Σ_D^γ and Σ_M^γ and the similarity matrix Φ_Ω^γ generated by the AGR sub-module to the object regional proposals with the same category labels by the mapping mechanism g. The purpose is to obtain the appearance geometric features $\Sigma_I^{R\gamma}$, $\Sigma_D^{R\gamma}$ and $\Sigma_M^{R\gamma}$ and the similarity matrix $\Phi_\Omega^{R\gamma}$ that can be used to assist in the detection of real samples, as shown in Figure 6b. The calculation formula is as follows:

$$g : \hbar(\Sigma_{ij}^\gamma, \Phi_\Omega^\gamma) \rightarrow \hbar^R(\Sigma_{ij}^{R\gamma}, \Phi_\Omega^{R\gamma}) \tag{4}$$

where $g^{-1}(\hbar^R) \subseteq \hbar$; g is a unidirectional surjection function; $\hbar$ and $\hbar^R$ correspond to the artificial sample feature space and the real sample feature space, respectively.

Finally, the appearance geometric features $\Sigma_I^{R\gamma}$, $\Sigma_D^{R\gamma}$ and $\Sigma_M^{R\gamma}$ and the similarity matrix $\Phi_\Omega^{R\gamma}$ are send to the Graph Neural Network (GNN) [28] G_a for reasoning the appearance geometric features X_a of the real samples, as shown in Figure 6c. Likewise, the normalized adjacency matrix $\widetilde{\mathbf{A}}_a$ of the G_a is composed of $\Phi_\Omega^{R\gamma}$; X_a^l and X_a^{l+1} correspond to the input and output features of the G_a respectively; $\mathbf{W}_a^l$ is the learned weight matrix; and σ represents a ReLU activation function. The calculation formula is as follows:

$$X_a^{l+1} = \sigma(\widetilde{\mathbf{A}}_a X_a^l \mathbf{W}_a^l) \tag{5}$$

2.4. Spatial Geometric Reasoning

In aerial images of insulator defects, the insulator defect area is relatively small compared to the whole image, which makes defect detection more difficult. Therefore, to greater extract the defect features, it is necessary to design a module to capture the interaction information between the defect area and the global area. The obvious spatial information includes: (1) the defect area must be located on the main body of the insulator strings; (2) there is also a spatial positional relationship between different defects; (3) the spatial positional relationship also exists between the defect area and the adjacent insulators. By extracting the spatial information and utilizing the Graph Neural Network (GNN) for learning, the spatial geometry features for assisting the localization and regression of region proposals in the object can be obtained, as shown in Figure 7.

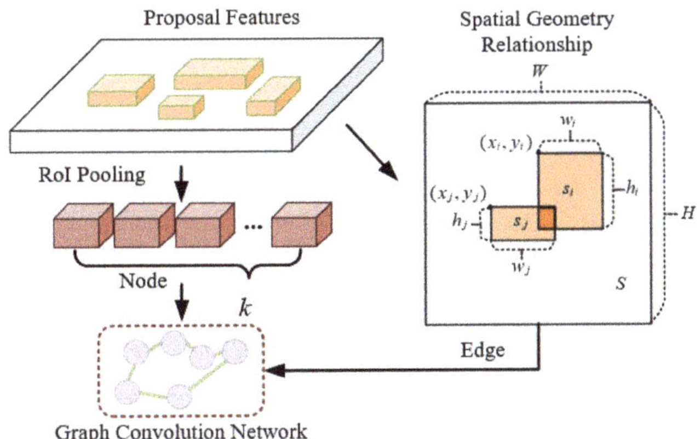

Figure 7. Architecture of the SGR model.

In detail, the proposal features were obtained from the real samples by RPN. Then it is submitted to RoI Pooling to generate the 128 regional proposals for a graph convolution network $G : G = (\vartheta, E)$. In the GNN, each node v corresponds to a region proposal. To reduce the redundant noise information in the modeling process of spatial geometric location information and make the output node information smooth, the k region proposals are selected as nodes θ of the GNN. where k is a hyperparameter; $\theta \in \vartheta$, ϑ is a set of nodes.

The spatial geometric relationship r_{ij} between the region proposals is represented by a set of spatial location calculation laws. This enables the interaction of information between the region proposals and overcomes the effects of different scales of defects and different spatial locations. Due to the few defect labels in a single image, the spatial location relationship of the regional proposals with different labels is simple. Thus, only little node information is required to construct a graph neural network. The calculation formula is as follows:

$$r_{ij} = \left\{ \frac{(x_i - x_j)}{w_j}, \frac{(y_i - y_j)}{h_j}, \frac{(x_i - x_j)^2}{w_j^2}, \frac{(y_i - y_j)^2}{h_j^2}, \log\left(\frac{w_i}{w_j}\right), \log\left(\frac{h_i}{h_j}\right) \right\} \tag{6}$$

where (x_i, y_i) is a center point coordinates of the region proposals; w_i and h_i correspond to the width and height of the region proposals, respectively.

The spatial location relationship r_{ij} is supplied to a ReLU activation function and then normalize it to obtain the final spatial geometry relationship e_{ij}. The spatial geometric relation e_{ij} as the node-to-node edge in the GNN. The calculation formula is as follows:

$$e_{ij} = \frac{exp(ReLU(W_s r_{ij}))}{\sum_k exp(ReLU(W_s r_{kj}))} \tag{7}$$

where $e_{ij} \in E \in \mathbb{R}^{k \times k}$; E is a set of edges; W_s is a learnable weight parameter.

After obtaining information about the nodes and edges used for the GNN, we construct a GNN G_s to represent the spatial geometric features of the defect location information. where $\tilde{A}_s$ is the normalized adjacency matrix formed by e_{ij}; X_s^l and X_s^{l+1} correspond to the input features and output features of the GNN, respectively; W_s^l is the learned weight matrix; σ represents the ReLU activation function. The calculation formula is as follows:

$$X_s^{l+1} = \sigma(\tilde{A}_s X_s^l W_s^l) \tag{8}$$

2.5. Training Method

This section describes the loss function, feature fusion approaches, and the model training process. The weighted sums operation is adopted to fuse the original features, the appearance of geometric features, and spatial geometric features. This can effectively avoid the impact of feature redundancy on the model performance. The calculation formula is as follows:

$$f_{all} = f + \lambda X_a + (1 - \lambda) X_s \tag{9}$$

where λ is a hyperparameter.

The loss function L_2 is a multi-task loss on each proposal RoI for jointly training classification and bounding-box regression. The calculation formula is as follows:

$$L_2(p_i, t_i) = \frac{1}{N_{cls}} \sum_i L_{cls}(p_i, p_i^*) + \mu \frac{1}{N_{reg}} \sum_i p_i^* L_{reg}(t_i, t_i^*) \tag{10}$$

where i is the index of proposals. The p_i^* and t_i^* are the ground-truth label and box, respectively. The classification loss L_{cls} is cross entropy loss over multiple classes of fittings. The regression loss L_{reg} is defined by $Smooth_{L_1}$ loss function. In practice, we adopt the generally used technique SGD to implement our algorithm.

The model training process is carried out with the following steps:

1. First, to obtain the deep features f of artificial defect samples, our method trains a classifier named Classifier-0 by cross-entropy loss L_1 in the preparatory phase;
2. Second, compute Σ_{ij}^γ and Φ_Ω^γ according to Equations (1) and (2), respectively.
3. Third, the Bboxs are obtained from the real samples through CNN, RPN, and RoI Pooling in turn by loss function L_2 in the formal phase;
4. Fourth, input the category information and feature information contained in Bboxs. Based on this information, X_a is computed by Equations (4) and (5); X_s is computed by Equations (6)–(8);
5. Fifth, fuse original features and enhanced features X_s and X_a together and output the f_{all};
6. Last, train GNN G_s and G_a for reasoning and learning spatial geometric feature X_s and appearance geometric feature X_a by loss function L_2, respectively;.

3. Experiment and Numerical Result

3.1. Datasets Description

AS-I Dataset: AS-I defect dataset is a high-quality database of different angles generated from 3D defect models, which was collected to solve the problem of insulator geometric defect detection. The dataset includes two categories of insulator defects, including 1175 damaged insulator images, 1814 missing insulator images, and 1360 defect-free insulator images. The resolution of each raw image is 1600×1200. All images are pure color backgrounds which makes the high contrast between object and background in the total dataset. Figure 4 shows a part of the AS-I dataset.

RS-I Dataset: The RS-I defect dataset consists of 332 images of two types of defects, insulator damage and insulator missing, including 303 insulator labels, 165 missing insulator labels, and 183 damaged insulator labels. These defect images have different resolutions and most of them contain a series of noises, such as the diversity of defect shapes, and relatively low contrast between the object and the background. All these factors pose great challenges to detection. Figure 1 shows a part of the RS-I dataset.

3.2. Implementation Details

Parameters Setting: We use Faster R-CNN as our baseline model and adopt a pretrained ResNet50 [29] on ImageNet as the backbone network. In RS-I and AS-I datasets training stage, we both adopt mini-batch stochastic gradient descent (SGD) optimizer with a momentum of 0.9 for network optimization. In RS-I and AS-I datasets, the training set and the validation set are divided in the ratio of 1:1 (i.e., 166 training/test samples) Detailed parameters setting is shown in Table 1.

Computation Platform: We implement our method on the PyCharm with the open-source toolbox PyTorch. We run our method in an NVIDIA GTX 1080Ti GPU on Ubuntu 16.04.

Table 1. Experimental basic parameters setting.

Parameters Setting	RS-I Dataset	AS-I Dataset
Backbone	Resnet50	Resnet34
Optimizer	SGD	SGD
Batch size	1	16
Epoch	50	100

3.3. Evaluation Metrics

In terms of evaluation metrics, binary precision (P) and recall (R) are chosen to validate the detection performance in our work. The binary P and R are calculated by:

$$P = \frac{TP}{TP + FP} \tag{11}$$

$$R = \frac{TP}{TP + FN} \tag{12}$$

where TP, FP and FN represent the number of true positive, false positive and false negative samples respectively.

Furthermore, average precision (AP) for binary pest localization is applied as a comprehensive evaluation metric to take the P and R into consideration together. In single object detection, the AP of detecting an object is computed by the integration of the precision-recall (PR) curve. The calculation formula is as follows:

$$AP = \int_0^1 PdR \tag{13}$$

In the multi-object detecting task, we usually select the mean average precision (mAP) that is obtained by taking an average of APs from all the fitting categories to evaluate the model accuracy. The calculation formula is as follows:

$$mAP = \frac{1}{C} \sum_{i=1}^{C} AP_i \tag{14}$$

For a comprehensive and comprehensive evaluation, we adopt the metrics from COCO detection evaluation criteria[30], i.e., mAP across different intersection over union (IoU) thresholds (IoU = 0.5:0.95, 0.5, 0.75). We also use average recall (AR) with a different number of given detections per image (1, 10).

3.4. Comprehensive Comparison and Analysis

For the sake of verifying the performance of the proposed method more comprehensively, The COCO evaluation criteria are employed to compare MGRN with other object detection models, viz. YOLOv3 [22], Retina Net [31], SSD [21], Cascade R-CNN [32], Libra R-CNN [33] and baseline model. The results are listed in Table 2.

It can be seen from Table 2 that the AP^{50} value of the MGRN model reached 54.7%, an increment of 4.9% compared with the baseline model. Compared with Libra R-CNN and Cascade R-CNN model, the AP^{50} value of our model was increased by 3.9% and 5.1%, respectively. This demonstrates that our method can improve both the accuracy and false discovery rate by extracting the appearance geometric features and spatial geometric features of defects. We also compare our method to other object detection models in Table 2. It can be observed that the MGRN outperforms other competing methods by a large margin. Compared with other models, MGRN has greatly improved the detection effect of hard-detection insulator defects. In terms of damaged defect detection, the proposed MGRN increases the AP^{50} by 4% compared with the baseline model. Similarly, the proposed MGRN increases the AP^{50} by 3.2% compared with the baseline model in the missing defect detection. The detection results of each category are shown in Figure 8.

Table 2. Performance comparison of different detection methods. Bolded numbers indicate optimal.

Method	$AP^{50:95}$	AP^{50}	AP^{75}	AR^1	AR^{10}
YOLOv3 [22]	8.4	23.3	5.2	15.7	24.8
Retina Net [31]	10.3	23.9	7.0	16.9	26.3
SSD [21]	18.0	41.4	13.8	23.1	30.2
Cascade R-CNN [32]	24.1	49.6	18.2	29.0	36.0
Libra R-CNN [33]	20.1	50.8	11.2	26.9	36.6
Baseline [24]	23.9	49.8	18.0	30.6	34.7
MGRN (ours)	**25.4**	**54.7**	**24.6**	**31.8**	**37.8**

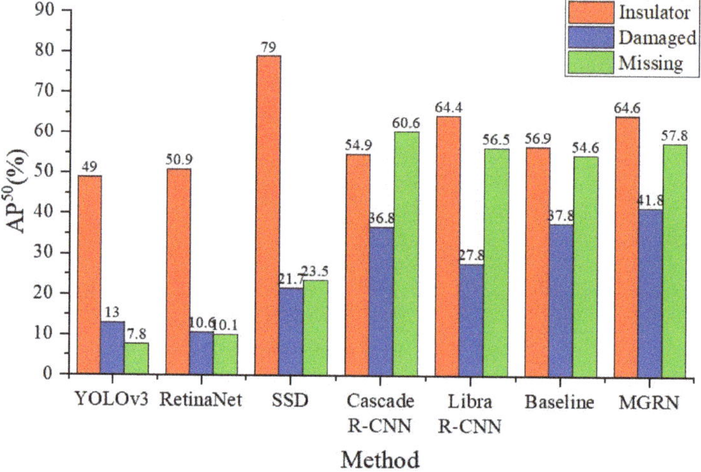

Figure 8. Comparison of defect detection results of various models.

To qualitatively analyze the detection effect of the proposed model, the visualization comparison between the baseline and MGRN is shown in Figure 9, where the blue boxes represent the repeated detection of insulator and insulator defects and the red boxes represent the missed-detected insulator defects. As shown in Figure 9, damaged and missing defects can be hardly detected due to the size of the defect in the image being too small.

In particular, the similarity in shape and background of defects and adjacent insulators results in missed detections in the baseline model. MGRN effectively utilizes the appearance geometric information and spatial geometric information of defects, reduces the influence of the texture color information on the model, and greatly improves the detection precision. As shown in the red box in Figure 9b, defects with minor features and similar background colors are not detected by the baseline model. However, the MGRN model overcomes this problem in Figure 9e, and it is able to detect this defect.

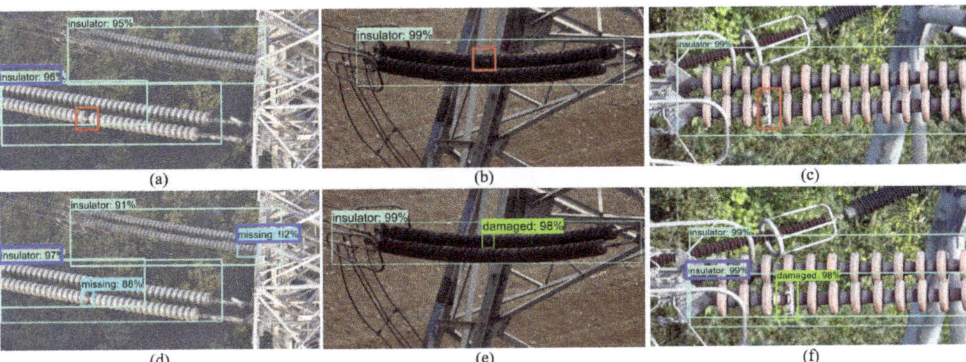

Figure 9. Qualitative result comparison for defects detection. (**a–c**) baseline detection results; (**d–f**) MGRN detection results.

3.5. Ablative Study

To evaluate the proposed solution, this article conducts a rank of ablative experiments, including hyperparameter setting in the total model, and the effects of the fusion feature. We did some tests on the Faster R-CNN baseline model. All the evaluations of these ablative experiments are based on the RS-I dataset.

Contributions of Each Sub-module: The proposed MGRN consists of three sub-modules, i.e., AGR sub-module, PFT sub-module, and SGR sub-module. The AGR sub-module and PFT sub-module are used to extract the appearance geometric features of defects. The SGR module is used to extract the spatial geometric features of defects. We compare the detection results of different sub-modules on the RS-I dataset. The result is shown in Table 3. The AGF represents the model of only using the AGR sub-module and the PFT sub-module for defect detection. The SGF represents the model of only using the SGR sub-module for defect detection. It can be observed that the performance will increase by 3.2% and 4.2% if we add either sub-network, which can further validate the effectiveness of each component in the proposed MGRN.

Table 3. Comparison of test results of different sub-modules. Bolded numbers indicate optimal.

Method	AP^{50}	AP^{75}	AR^1	AR^{10}
Baseline	49.8	18.0	30.6	34.7
Baseline+SGF	53.1	22.9	30.3	36.7
Baseline+AGF	54.0	21.1	**32.5**	36.7
MGRN (ours)	**54.7**	**24.6**	31.8	**37.8**

Impact of Different Regional Proposal Parameters: we compare the detection performance with the different number of region proposals based on the SGR sub-module. The hyperparameter $k \in \{2, 4, 8, 16, 32, 64, 128\}$. Table 4 shows the AP^{50} result with different k, marking best and second best. It can be observed in Table 4 that the SGR sub-module can preferably improve the baseline model when only the four region proposals are used to represent the spatial geometric features. At this point, the AP^{50} value of the MGRN model with only the SGR sub-module reached 53.1%. It can be observed that as the value of k increases, the performance of the model declines instead. When all 128 region proposals are used to structure the network, the SGR sub-module accomplishes the maximum results in AP^{50}, AP^{75}, AR^1, and AR^{10}. However, this will generate redundant information and a huge computational overhead. Thus, we choose the $k = 4$ in the SGR sub-module.

Table 4. Comparison of test results of different k. The underlined numbers indicate sub-optimal, the bolded numbers indicate optimal.

k	AP^{50}	AP^{75}	AR^1	AR^{10}
2	51.8	22.4	30.9	35.6
4	**53.1**	<u>22.9</u>	30.3	<u>36.7</u>
8	50.0	18.5	30.3	34.0
16	50.2	18.7	28.6	33.5
32	50.6	18.7	29.4	34.3
64	51.7	14.9	30.3	36.0
128	51.5	**25.8**	**32.7**	**37.7**

Impact of Different Similarity Matrix with Different Accuracy Levels: We use different accuracy to conduct the ablative experiments from 95% to 99%, the accuracy $\gamma \in \{95\%, 96\%, 97\%, 98\%, 99\%\}$ Classifiers with different accuracy determine the results of appearance geometric feature extraction for artificial samples. We calculated the similarity matrix generated in the AGC sub-module with different precision, as shown in Figure 10. It can be observed in Figure 10 that S_{I-D} is greater than S_{I-M} and is greater than S_{D-M} at different accuracy.

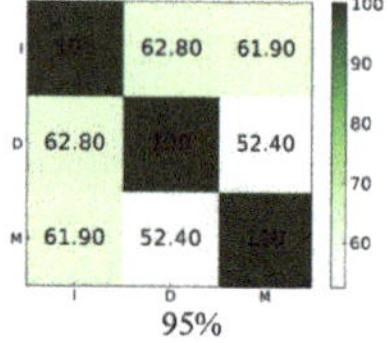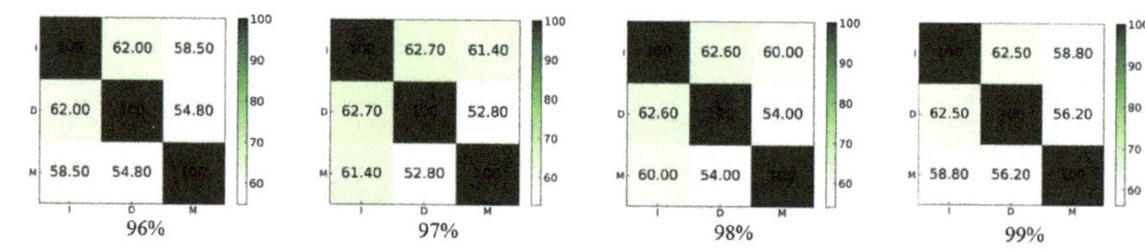

Figure 10. The similarity matrix with different accuracy.

To prove the effectiveness of appearance geometric feature generated by AGC sub-module and PFT sub-module with different accuracy of the similarity matrix. The results are shown in Table 5. The AP^{50} value of the model gradually increases as the accuracy γ improves. When the accuracy γ is 99%, the AP^{50} value of the MGRN model with only the AGR sub-module and PFT sub-module reached 54%. This shows that the appearance of geometric features from the artificial samples can be effectively expressed through the graph reasoning network G_a, thereby improving the insulator defects detection performance.

Table 5. Comparison of test results of different accuracy.

Accuracy γ	AP^{50}	AP^{75}	AR^1	AR^{10}
95%	50.0	19.3	32.5	37.4
96%	51.1	18.0	31.4	37.6
97%	52.4	21.4	31.9	37.2
98%	52.9	22.2	32.1	36.0
99%	54.0	21.1	32.5	36.7

Impact of Different Features Fusion Scales: This work evaluates the effect of different feature fusion scales on the MGRN model, and the corresponding performance is shown in Figure 11. It can be observed from Figure 11 that the MGRN model displays better and reaches 54.7% when the hyperparameter λ equals 0.4, which demonstrates the feature maps fusion can significantly promotes the performance of the detection in the proposed solution. Figure 11 shows that the AGR sub-module, PFT sub-module, and SGR sub-module can effectively solve the problem of insulator geometric defect detection with strong robustness. In addition, the large gap in feature fusion rate will reduce the detection performance of the MGRN model as shown in Figure 11. However, the overall performance of the model still improves over the baseline model.

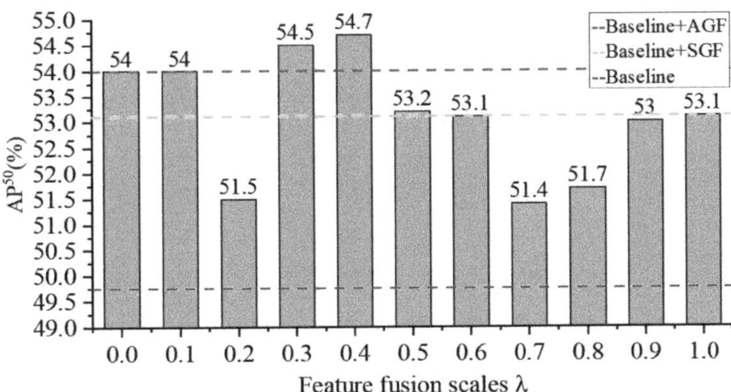

Figure 11. Performance of different feature fusion scales.

4. Conclusive Remarks

This paper developed an automatic reasoning detection network based on multi-geometric features for defect detection of electric transmission line insulators. In the proposed solution, the AGR sub-module is developed to extract the appearance geometric features of defects from the artificial insulator samples with defects. Through designing the PFT sub-module, the extracted appearance geometric features are made available to the graph convolutional network for reasoning learning. In addition, the SGR sub-module is developed to identify the space geometric position relationship between defects to capture the interaction information under the regional proposals.

The proposed solution is extensively assessed through experiments against the existing solutions and the numerical results demonstrated that the proposed MGRN-based solution significantly advanced the benchmarking solutions on insulator geometric defect detection with limited data availability, and the AP^{50} improvement of the scarcity sample is up to 41.8%. To our knowledge, it is the first work that artificial samples are expressed in deep space and then transformed into real samples and applied to transmission line insulator geometric defect detection. In future research, we plan to seek a better appearance geometric representation method of defects to improve the detection performance and meet the requirements of real-world scenarios.

Author Contributions: Conceptualization, Z.H. and Q.W.; methodology, Z.H. and Q.W.; software, Z.H.; validation, Y.Z., Q.Y. and K.Y.; formal analysis, K.Y.; investigation, Z.H.; resources, Y.Z.; data curation, Y.Z.; writing—original draft preparation, Z.H.; writing—review and editing, Z.H. and Q.Y.; visualization, Z.H.; supervision, Y.Z.; project administration, K.Y.; funding acquisition, Y.Z. All authors have read and agreed to the published version of the manuscript.

Funding: This work is supported in part by the National Natural Science Foundation of China (NSFC) under grant number U21A20486, 61871182, by the Natural Science Foundation of Hebei Province of China under grant number F2021502008.

Institutional Review Board Statement: Not applicable.

Informed Consent Statement: Not applicable.

Data Availability Statement: Not applicable.

Acknowledgments: Special thanks are given to the Department of Automation and North China Electric Power University and Smart Energy Systems (SES) Laboratory, College of Electrical Engineering, Zhejiang University.

Conflicts of Interest: The authors declare no conflict of interest. The funders had no role in the design of the study; in the collection, analysis, or interpretation of data; in the writing of the manuscript; or in the decision to publish the results.

Abbreviations

The following abbreviations are used in this manuscript:

MGRN	Multi-Geometric Reasoning Network
SGR	Spatial Geometric Reasoning
AGR	Appearance Geometric Reasoning
PFT	Parallel Feature Transformation
SSD	Single Shot MultiBox Detector
CNN	Convolutional Neural Network
YOLO	You Only Look Once
RPN	Regional Proposal Network
GPU	Graphics Processing Unit
P	Precision
R	Recall
TP	True Positive
FP	False Positive
FN	False Negative
UAVs	Unmanned Aerial Vehicles

References

1. Li, S.; Li, J. Condition monitoring and diagnosis of power equipment: Review and prospective. *High Volt.* **2017**, *2*, 82–91. [CrossRef]
2. Park, K.C.; Motai, Y.; Yoon, J.R. Acoustic fault detection technique for high-power insulators. *IEEE Trans. Ind. Electron.* **2017**, *64*, 9699–9708. [CrossRef]
3. Lei, X.; Sui, Z. Intelligent fault detection of high voltage line based on the Faster R-CNN. *Measurement* **2019**, *138*, 379–385. [CrossRef]
4. Choi, I.H.; Koo, J.B.; Son, J.A.; Yi, J.S.; Yoon, Y.G.; Oh, T.K. Development of equipment and application of machine learning techniques using frequency response data for cap damage detection of porcelain insulators. *Appl. Sci.* **2020**, *10*, 2820. [CrossRef]
5. Han, J.; Yang, Z.; Xu, H.; Hu, G.; Zhang, C.; Li, H.; Lai, S.; Zeng, H. Search like an eagle: A cascaded model for insulator missing faults detection in aerial images. *Energies* **2020**, *13*, 713. [CrossRef]
6. Nguyen, V.N.; Jenssen, R.; Roverso, D. Automatic autonomous vision-based power line inspection: A review of current status and the potential role of deep learning. *Int. J. Electr. Power Energy Syst.* **2018**, *99*, 107–120. [CrossRef]
7. Tao, G.; Chen, F.; Wei, W.; Ping, S.; Lei, S.; Tianzhu, C. Electric insulator detection of UAV images based on depth learning. In Proceedings of the 2017 2nd International Conference on Power and Renewable Energy (ICPRE), Chengdu, China, 20–23 September 2017; pp. 37–41.
8. Ma, Y.; Li, Q.; Chu, L.; Zhou, Y.; Xu, C. Real-time detection and spatial localization of insulators for UAV inspection based on binocular stereo vision. *Remote Sens.* **2021**, *13*, 230. [CrossRef]
9. Ni, L.; Ma, Y.; Lin, Q.; Yang, J.; Jin, L. Research on Insulator Defect Detection Method Based on Image Processing and Watershed Algorithm. In Proceedings of the 2021 International Conference on Advanced Electrical Equipment and Reliable Operation (AEERO), Beijing, China, 15–17 October 2021; pp. 1–6.
10. Murthy, V.S.; Tarakanath, K.; Mohanta, D.; Gupta, S. Insulator condition analysis for overhead distribution lines using combined wavelet support vector machine (SVM). *IEEE Trans. Dielectr. Electr. Insul.* **2010**, *17*, 89–99. [CrossRef]
11. Liao, S.; An, J. A robust insulator detection algorithm based on local features and spatial orders for aerial images. *IEEE Geosci. Remote Sens. Lett.* **2014**, *12*, 963–967. [CrossRef]
12. Zhang, Q.; Li, W.; Li, H.; Wang, J. Self-blast state detection of glass insulators based on stochastic configuration networks and a feedback transfer learning mechanism. *Inf. Sci.* **2020**, *522*, 259–274. [CrossRef]
13. Tao, X.; Zhang, D.; Wang, Z.; Liu, X.; Zhang, H.; Xu, D. Detection of power line insulator defects using aerial images analyzed with convolutional neural networks. *IEEE Trans. Syst. Man Cybern. Syst.* **2018**, *50*, 1486–1498. [CrossRef]
14. Liu, J.; Liu, C.; Wu, Y.; Xu, H.; Sun, Z. An Improved Method Based on Deep Learning for Insulator Fault Detection in Diverse Aerial Images. *Energies* **2021**, *14*, 4365. [CrossRef]
15. Kalla, U.K.; Suthar, R.; Sharma, K.; Singh, B.; Ghotia, J. Power quality investigation in ceramic insulator. *IEEE Trans. Ind. Appl.* **2017**, *54*, 121–134. [CrossRef]
16. Gonçalves, R.S.; Agostini, G.S.; Bianchi, R.A.; Homma, R.Z.; Sudbrack, D.E.T.; Trautmann, P.V.; Clasen, B.C. Inspection of Power Line Insulators: State of the Art, Challenges, and Open Issues. In *Handbook of Research on New Investigations in Artificial Life, AI, and Machine Learning*; IGI Global: Hershey, PA, USA, 2022; pp. 462–491.
17. She, L.; Fan, Y.; Wang, J.; Cai, L.; Xue, J.; Xu, M. Insulator Surface Breakage Recognition Based on Multiscale Residual Neural Network. *IEEE Trans. Instrum. Meas.* **2021**, *70*, 1–9. [CrossRef]
18. Sadykova, D.; Pernebayeva, D.; Bagheri, M.; James, A. IN-YOLO: Real-time detection of outdoor high voltage insulators using UAV imaging. *IEEE Trans. Power Deliv.* **2019**, *35*, 1599–1601. [CrossRef]

9. Wang, S.; Liu, Y.; Qing, Y.; Wang, C.; Lan, T.; Yao, R. Detection of insulator defects with improved resnest and region proposal network. *IEEE Access* **2020**, *8*, 184841–184850. [CrossRef]
0. Zhang, H.; Wu, C.; Zhang, Z.; Zhu, Y.; Lin, H.; Zhang, Z.; Sun, Y.; He, T.; Mueller, J.; Manmatha, R.; et al. Resnest: Split-attention networks. In Proceedings of the IEEE/CVF Conference on Computer Vision and Pattern Recognition, New Orleans, LA, USA, 21–24 June 2022; pp. 2736–2746.
1. Liu, W.; Anguelov, D.; Erhan, D.; Szegedy, C.; Reed, S.; Fu, C.Y.; Berg, A.C. Ssd: Single shot multibox detector. In Proceedings of the European Conference on Computer Vision, Amsterdam, The Netherlands, 11–14 October 2016; Springer: Berlin/Heidelberg, Germany, 2016; pp. 21–37.
2. Redmon, J.; Farhadi, A. Yolov3: An incremental improvement. *arXiv* **2018**, arXiv:1804.02767.
3. Girshick, R. Fast r-cnn. In Proceedings of the IEEE International Conference on Computer Vision, Santiago, Chile, 7–13 December 2015; pp. 1440–1448.
4. Ren, S.; He, K.; Girshick, R.; Sun, J. Faster r-cnn: Towards real-time object detection with region proposal networks. *Adv. Neural Inf. Process. Syst.* **2015**, *28*, 91–99. [CrossRef]
5. He, K.; Gkioxari, G.; Dollár, P.; Girshick, R. Mask r-cnn. In Proceedings of the IEEE International Conference on Computer Vision, Venice, Italy, 22–29 October 2017; pp. 2961–2969.
6. Zhai, Y.; Wang, Q.; Yang, X.; Zhao, Z.; Zhao, W. Multi-fitting Detection on Transmission Line based on Cascade Reasoning Graph Network. *IEEE Trans. Power Deliv.* **2022**. [CrossRef]
7. Kang, G.; Gao, S.; Yu, L.; Zhang, D. Deep architecture for high-speed railway insulator surface defect detection: Denoising autoencoder with multitask learning. *IEEE Trans. Instrum. Meas.* **2018**, *68*, 2679–2690. [CrossRef]
8. Zhou, J.; Cui, G.; Hu, S.; Zhang, Z.; Yang, C.; Liu, Z.; Wang, L.; Li, C.; Sun, M. Graph neural networks: A review of methods and applications. *AI Open* **2020**, *1*, 57–81. [CrossRef]
9. He, K.; Zhang, X.; Ren, S.; Sun, J. Deep residual learning for image recognition. In Proceedings of the IEEE Conference on Computer Vision and Pattern Recognition, Las Vegas, NV, USA, 27–30 June 2016; pp. 770–778.
0. Lin, T.Y.; Maire, M.; Belongie, S.; Hays, J.; Perona, P.; Ramanan, D.; Dollár, P.; Zitnick, C.L. Microsoft coco: Common objects in context. In Proceedings of the European Conference on Computer Vision, Zurich, Switzerland, 6–12 September 2014; Springer: Berlin/Heidelberg, Germany, 2014; pp. 740–755.
1. Lin, T.Y.; Goyal, P.; Girshick, R.; He, K.; Dollár, P. Focal loss for dense object detection. *IEEE Int. Conf. Comput. Vis.* **2017**, *42*, 2980–2988.
2. Cai, Z.; Vasconcelos, N. Cascade R-CNN: High quality object detection and instance segmentation. *IEEE Trans. Pattern Anal. Mach. Intell.* **2019**, *43*, 1483–1498. [CrossRef] [PubMed]
3. Pang, J.; Chen, K.; Shi, J.; Feng, H.; Ouyang, W.; Lin, D. Libra r-cnn: Towards balanced learning for object detection. In Proceedings of the IEEE/CVF Conference on Computer Vision and Pattern Recognition, Long Beach, CA, USA, 15–20 June 2019; pp. 821–830.

Article

Insulator Umbrella Disc Shedding Detection in Foggy Weather

Rui Xin [1], Xi Chen [1], Junying Wu [1], Ke Yang [2,*], Xinying Wang [2] and Yongjie Zhai [2]

[1] State Grid Hebei Information and Telecommunication Branch, Shijiazhuang 050013, China; xinruicn@163.com (R.X.); m18830267026@163.com (X.C.); kkyy237@163.com (J.W.)

[2] Department of Automation, North China Electric Power University, Baoding 071003, China; wangxinying@ncepu.edu.cn (X.W.); zhaiyongjie@ncepu.edu.cn (Y.Z.)

* Correspondence: yangke@ncepu.edu.cn; Tel.: +86-158-3221-7986

Abstract: The detection of insulator umbrella disc shedding is very important to the stable operation of a transmission line. In order to accomplish the accurate detection of the insulator umbrella disc shedding in foggy weather, a two-stage detection model combined with a defogging algorithm is proposed. In the dehazing stage of insulator images, solving the problem of real hazy image data is difficult; the foggy images are dehazed by the method of synthetic foggy images training and real foggy images fine-tuning. In the detection stage of umbrella disc shedding, a small object detection algorithm named FA-SSD is proposed to solve the problem of the umbrella disc shedding occupying only a small proportion of an aerial image. On the one hand, the shallow feature information and deep feature information are fused to improve the feature extraction ability of small targets; on the other hand, the attention mechanism is introduced to strengthen the feature extraction network's attention to the details of small targets and improve the model's ability to detect the umbrella disc shedding. The experimental results show that our model can accurately detect the insulator umbrella disc shedding defect in the foggy image; the accuracy of the defect detection is 0.925, and the recall is 0.841. Compared with the original model, it improved by 5.9% and 8.6%, respectively.

Keywords: insulator umbrella disc shedding; defect detection; dehazing algorithm; feature fusion; attention mechanism

Citation: Xin, R.; Chen, X.; Wu, J.; Yang, K.; Wang, X.; Zhai, Y. Insulator Umbrella Disc Shedding Detection in Foggy Weather. *Sensors* **2022**, *22*, 4871. https://doi.org/10.3390/s22134871

Academic Editor: Paweł Pławiak

Received: 18 May 2022
Accepted: 25 June 2022
Published: 28 June 2022

1. Introduction

Transmission lines are an important part of the power grid and are crucial to the safe and stable operation of the power grid [1,2]. In transmission lines, insulators are the basic equipment used for electrical isolation and mechanical fixation in high-voltage transmission systems [3,4]. Since the insulators remain exposed, environmental factors inevitably cause damage to them, and the resulting insulator failures can seriously affect the safe and stable operation of the power grid [5,6]. Therefore, timely detection of insulator defects and early treatment can effectively reduce the occurrence of insulator failures [7,8]. The defects of insulators mainly include umbrella disc shedding, umbrella disc damage, dirt, and icing. Among these defects, insulator umbrella disc shedding is the most common, the most numerous, and the most harmful defect. With the rapid development of 5G technology and AI technology [9,10], combined with 5G high-speed data transmission and target detection technology, through the all-weather monitoring of transmission lines, insulator defects can be found in time, effectively reducing the transmission line failures caused by insulator defects [11]. The advantage of 5G technology is that it can achieve high-speed data transmission, which can not only ensure image quality but also ensure real-time detection. Transmission line insulator defect detection based on 5G and AI is shown in Figure 1. First, HD cameras take videos of insulators; second, the captured data are compressed and transmitted to the monitoring center through 5G communication; third, stsffs decompress the data, process the video frame by frame, and use the corresponding defect detection model to detect and judge the defect level; finally, technicians take corrective measures, according to the defect level.

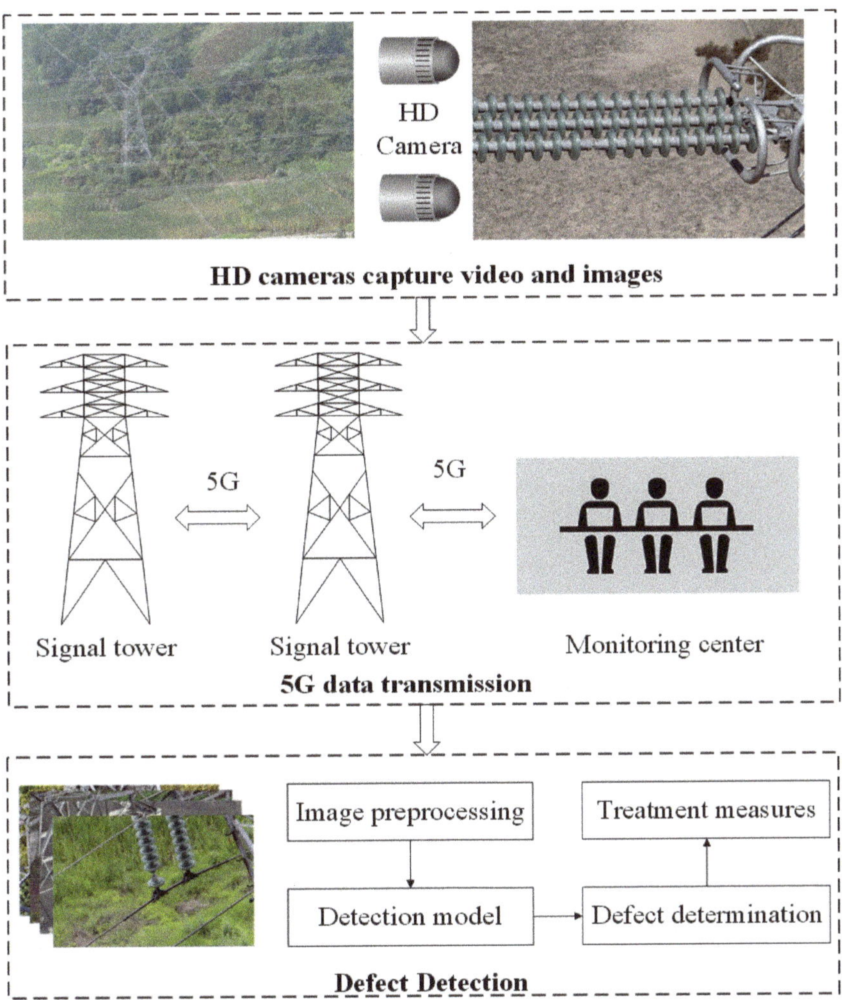

Figure 1. Insulator defect detection based on 5G and AI.

At present, the mainstream dehazing algorithms mainly include the dehazing algorithm based on image enhancement, the dehazing algorithm based on image restoration, and the dehazing algorithm based on CNN. The first method uses image processing to highlight image details and enhance contrast to make foggy images clearer. The specific algorithms include histogram equalization [12], wavelet transform [13], and the Retinex algorithm [14]. The second method is based on the physical model of atmospheric scattering, which can obtain the mapping relationship between the foggy image and the fog-free image; and it restores the foggy image to a clear image. The most representative algorithm is the dark channel prior dehazing algorithm proposed by He [15]. However, physical priors are not always reliable, and these priors do not apply to all hazy images, which makes the dehazing effect uncertain. The third method builds an end-to-end model through CNN to recover clear images from hazy images [16,17]. Such methods overcome the disadvantage of using physical priors; they are more efficient and perform better than traditional prior-based algorithms. Zhao [18] proposed a novel end-to-end convolutional neural network called the attention enhanced serial Unet++ [19] dehazing network (AESUnet) for single image dehazing, and the serial Unet++ module generated more realistic images with less color distortion. Gao [20] proposed an image dehazing model built with a convolutional neural network and Transformer to improve the quality of the restored image. However, CNN requires a large number of hazy and clear image pairs for training, which are difficult to obtain. Due to the lack of real foggy image datasets, many studies are carried out on synthetic foggy images, which makes it difficult to achieve good results when the dehazing algorithms are applied to real foggy images.

Researchers have investigated insulator defect detection. Zhang [21] proposed an optical image detection method based on deep learning and morphological detection. First of all, the Faster RCNN was used to locate the insulator and extract its target image from the detection image. Second, a segmentation method of the insulator image was proposed to remove the background of the target image. Finally, a mathematical model was established in the binary image to describe the defect of the insulator. Tao [22] proposed a novel deep CNN cascaded architecture to perform localization and detection of defects in insulators. The cascaded network transformed defect detection into a two-level object detection problem, which used a region proposal network-based CNN. The method first detected the insulator in the aerial image and then detected the shedding defect of the insulator umbrella disk on this basis. She [23] proposed a multiscale residual neural network for insulator surface damage identification, using three convolution kernels of different sizes to perform convolution filtering and feature map fusion to enrich the spatial correlation and channel correlation of feature maps. Aiming at the small proportion of the insulator umbrella disc shedding fault area in the entire image and the difficulty in detection, Zahng [24] introduced the densely connected feature pyramid network into the YOLOV3 [25] model to achieve high detection accuracy. Zhao [26] combined Faster R-CNN [27] and an improved FPN [28] to detect two types of insulator defects. However, the above studies were all to detect insulator defects under normal weather conditions. Under real environmental conditions, one will inevitably encounter complex weather conditions [29–31]. Foggy weather is the most common complex weather. Achieving the detection of insulator defects in foggy conditions is crucial for all-weather real-time monitoring of transmission lines. As shown in Figure 2, there is a clear difference between the insulator images in foggy and fog-free weather conditions.

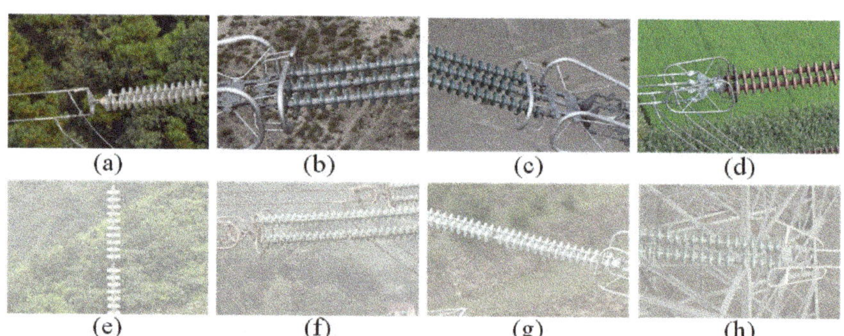

Figure 2. Images of insulators in foggy and fog-free weather conditions. (**a–d**) Clear images. (**e–h**) Foggy images.

This paper proposes a detection method for insulator umbrella disc shedding in foggy weather conditions. The main contributions of this paper are as follows:

(1) For the first time, the detection of insulator umbrella disc shedding in foggy conditions is realized, which provides a new way to detect transmission line defects in complex weather.

(2) A dehazing model with synthetic image pre-training and real image fine-tuning is proposed to solve the problem of the poor dehazing effect on real hazy images.

(3)   The FA-SSD model [32] is proposed to improve the accuracy and recall rate of insulator umbrella disc shedding detection.

2. Materials and Methods

As shown in Figure 3, the overall process of umbrella disc shedding detection included three parts: pre-training and fine-tuning of the defogging model, training with the clear insulator image datasets, and testing with the fogged insulator images.

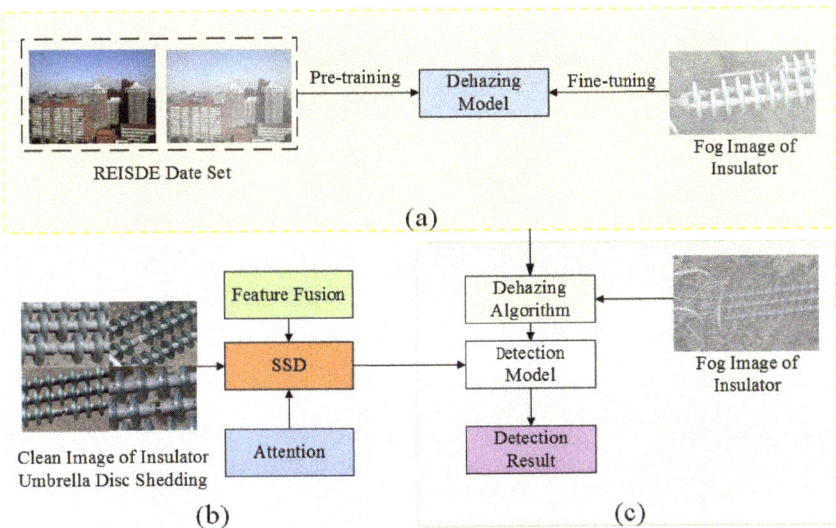

Figure 3. The overall process of umbrella disc shedding detection. (**a**) Dehaze model. (**b**) Training phase. (**c**) Testing phase.

The dehazing model was trained by synthetic foggy images, and the insulators with foggy images were fine-tuned to improve the dehazing effect of the algorithm. A feature fusion module and an attention module were added to the umbrella disc shedding detection model to improve the detection accuracy. In the detection of the insulator umbrella disc shedding, clear images of insulators were used for training, and images of insulators with fog were used for testing.

2.1. Dehazing Model

Inspired by the dehazing algorithm proposed by Chen [33], this paper adopted the method of pre-training and fine-tuning to improve the dehazing effect of the dehazing model. The training of the model was divided into two steps. The first step used a large number of haze-free images and artificially-generated fogged images from the REISDE dataset [34] to train the dehazing model, and the second step used the foggy insulator images to fine-tune the dehazing model to improve the dehazing ability of the dehazing model on fogged insulator images. During fine-tuning, physical priors were guided through the loss function. As shown in Figure 4, the dehazing model had a two-stage framework.

In the pre-training stage, an advanced dehazing model was adopted as the backbone. The pre-training phase used synthetic data for training, resulting in a pre-trained model on the synthetic domain. In the fine-tuning stage, the fog-free image J, transmission map t, and atmospheric light A were obtained through the backbone network. At the same time, three priors, including a dark channel prior, a bright channel prior, and the Contrast Limited Adaptive Histogram Equalization (CLAHE) were introduced, and the model was guided in the form of loss function.

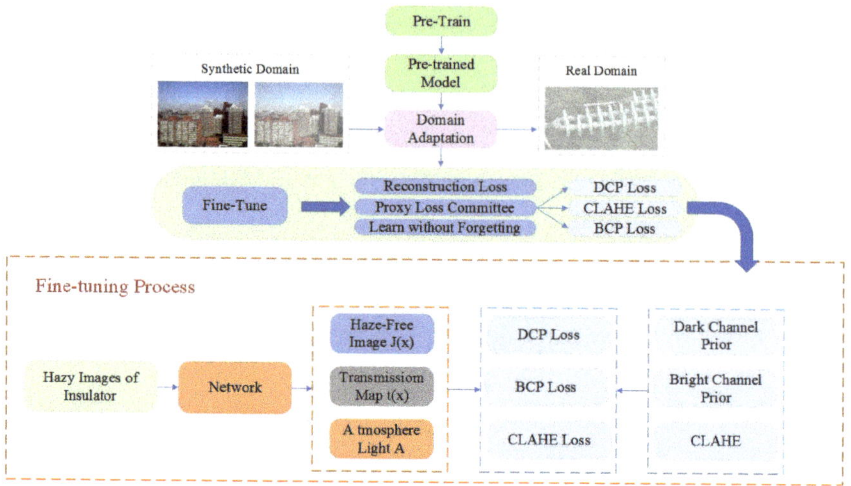

Figure 4. Structure of the dehaze model.

The loss function of the dark channel prior is shown as follows:

$$L_{DCP} = E(t, \tilde{t}) = t^T L t + \lambda (t - \tilde{t})^T (t - \tilde{t}) \tag{1}$$

where t and $\tilde{t}$ denote the transmission estimates from the DCP and the backbone network, respectively. L is a Laplacian-like matrix.

The loss function of the bright channel prior is shown as follows:

$$L_{BCP} = \left\| t - \tilde{t} \right\|_1 \tag{2}$$

where t and $\tilde{t}$ represent the transmission estimates from the BCP and the backbone network, respectively.

The loss function of the CLAHE reconstruction is shown as follows:

$$L_{CLAHE} = \| I - I_{CLAHE} \|_1 \tag{3}$$

where I is the original hazy input, and I_{CLAHE} is the reconstruction result by $J_{CLAHE}, \tilde{t},$ and $\tilde{A}$.

The role of the three loss functions is different. Dark channel prior greatly advances the model performance on real hazy images, bright channel prior helps make the resulting images brighter and with enhanced contrast, and CLAHE is used to achieve a balance between L_DCP and $LBCP$.

The total loss of the fine-tuning process was obtained by combining the three losses as follows:

$$L_{com} = \lambda_d L_{DCP} + \lambda_b L_{BCP} + \lambda_c L_{CLAHE} \tag{4}$$

where λ_d, λ_b, and λ_c are the tradeoff weights.

2.2. Fa-Ssd Model

Target detection includes target recognition and localization. For CNN, the two are contradictory [35]. Generally speaking, deep feature maps contain more semantic information, which is good for object recognition but not good for object localization; the difference is that the shallow feature map contains more detailed features, which is good for object localization but not good for object recognition. As shown in Figure 5, the SSD model adopts a feature pyramid structure to detect objects of different scales; small objects are detected on the shallow feature maps, and large objects are detected on the deep feature maps.

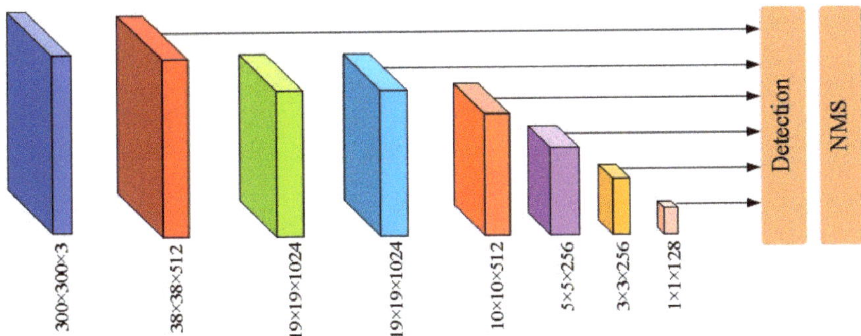

Figure 5. Structure of the SSD model.

However, the problem with this method is that the small target features generated by the shallow layer lack sufficient semantic information, and the detection of small targets still is not effective. In order to improve the detection ability of the SSD model for the insulator umbrella disk shedding, the FA-SSD model is proposed. As shown in Figure 6, the FA-SSD model adds a feature fusion module and an attention module to the SSD model.

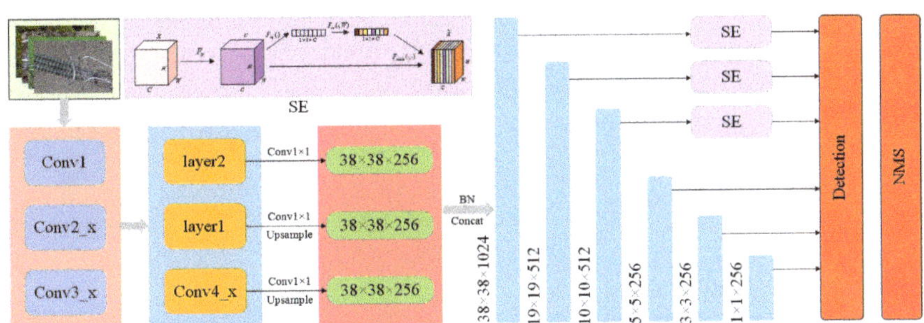

Figure 6. Structure of the FA-SSD model.

First, the insulator images were sent to the ResNet50 [36] feature extraction network to extract the features. Since the shallow feature maps contain richer small target detail information, Conv4_x in ResNet50 and two auxiliary convolutional layers were selected for feature fusion. The feature dimension of Conv4_x was $38 \times 38 \times 1024$, and the feature dimensions of the two auxiliary convolutional layers were $19 \times 19 \times 512$ and $10 \times 10 \times 512$. Then, in order to fuse the features of the three different scales simply and efficiently, the two auxiliary convolutional layers were upsampled using bilinear interpolation to make them the same size as Conv4_x. Finally, the feature map was concatenated and normalized to generate a new feature pyramid structure for the identification and localization of umbrella disc shedding. The parameters of each layer in the structure are shown in Table 1.

On this basis, in order to enhance the network's ability to extract low-level detail features, the SE channel attention module [37] was added to the lowest three layers of the feature pyramid.

Table 1. Input and output dimensions of each layer.

Layer Name	Input	Output
Conv1	$300 \times 300 \times 3$	$150 \times 150 \times 64$
Conv2_x	$150 \times 150 \times 64$	$75 \times 75 \times 256$
Conv3_x	$75 \times 75 \times 256$	$38 \times 38 \times 512$
Conv4_x	$38 \times 38 \times 512$	$38 \times 38 \times 1024$
layer1	$38 \times 38 \times 1024$	$19 \times 19 \times 512$
layer2	$19 \times 19 \times 512$	$10 \times 10 \times 512$

SE Module

The SE learns a set of weight coefficients through a small fully connected network to weigh each channel of the original feature map. In this way, different weights are assigned to each channel to enhance the feature extraction capability of the network. The implementation process of the SE was as follows:

(1) We performed convolution pooling and other operations on the input image to obtain a feature map:

$$u_c = v_c * X = \sum_{s=1}^{c'} v_c^s * x^s \tag{5}$$

where v_c and X represent the convolution kernel and the input image, respectively; v_c^s and x^s represent the convolution kernel and the sth channel of the input image, respectively; and c' represents the number of channels.

(2) We squeezed and compressed the feature map into one-dimensional features:

$$z_c = F_{sq}(u_c) = \frac{1}{H \times W} \sum_{i=1}^{H} \sum_{j=1}^{W} u_c(i,j) \tag{6}$$

where H and W represent the width and height of the feature map, respectively.

(3) For excitation, we performed activation operations on multiple channels to extract different features:

$$s = F_{ex}(z, W) = \sigma(g(z, W)) = \sigma(W_2 \delta(W_1 z)) \tag{7}$$

(4) We multiplied the obtained weight factor with the corresponding channel feature to obtain a new feature map:

$$\tilde{x}_c = F_{scale}(u_c, s_c) = s_c \cdot u_c. \tag{8}$$

3. Results

3.1. Experimental Environment

The proposed model used an NVIDIA RTX 2080Ti GPU for training and testing and the Ubuntu 18.04 LTS as the operating system; the training process was accelerated by CUDA 10.1; the computer language was Python 3.6, and the network framework was PyTorch. The batch size was set to 8, the learning rate was 0.003, the preprocessed size of the input image was 300×300, and the maximum number of iterations was 7800. The SSD was chosen as the baseline for improvement and comparison purposes.

The datasets used in the dehazing stage included the REISDE dataset and images of fogged insulators. The insulator images used in this paper were the aerial images of transmission line inspection, which were obtained by UAV. The datasets used in the object detection stage consisted of fogged insulator images, as well as fog-free insulator images. Since the insulators were in normal working condition most of the time, the defect images occupied a small proportion of the obtained aerial images. In addition, due to factors such as shooting environment, shooting angle, shooting distance, etc., many images were of poor quality. By cooperating with several power grid companies, we obtained some samples of insulator umbrella disk shedding. Among them, there were 160 images (the number of the insulator umbrella disc shedding was 176) with fog and 480 images (the number of the insulator umbrella disc shedding was 518) without fog. We used the images without fog as the training set and the images with fog as the test set. As shown in Figure 7, the insulator datasets contained glass insulators and ceramic insulators.

(a)   (b)

Figure 7. Glass insulators and ceramic insulators. (**a**) Glass insulators. (**b**) Ceramic insulators.

To compare the different models, precision(P), recall(R), and F_1 were used as model evaluation metrics. The higher the value, the better the detection performance of the model.

$$P = \frac{T_P}{T_P + F_P} \tag{9}$$

$$R = \frac{T_P}{T_P + F_N} \tag{10}$$

$$F_1 = \frac{2 \times P \times R}{P + R} \tag{11}$$

where TP and FP denote the number of correctly and incorrectly located defects, respectively. $TP + FP$ is the total number of located defects, and $TP + FN$ is the total number of actual defects. F_1 is the harmonic mean of precision and recall.

3.2. Ablation Experiment of Fa-Ssd Model

In order to verify the effectiveness of the feature fusion module and the attention module, the experiments were conducted on the original SSD model, the SSD model with the feature fusion module, the SSD model with the attention module, and the FA-SSD model. The visualization results of the FA-SSD model and the SSD model are shown in Figure 8.

(a)   (b)

Figure 8. Visualization of SSD and FA-SSD. (**a**) SSD. (**b**) FA-SSD.

In the experiment, the other parameters of the model training were guaranteed to be the same, and the obtained detection results are shown in Table 2.

Table 2. Results of the ablation experiment.

SSD	Feature Fusion	Attention	P	R	F_1
✓			0.866	0.755	0.806
✓	✓		0.899	0.769	0.828
✓		✓	0.877	0.793	0.832
✓	✓	✓	**0.909**	**0.817**	**0.860**

The detection performance of the FA-SSD was better than the methods that only added the feature fusion module or the attention module. Compared with the original SSD model, the accuracy rate was improved, the recall rate was improved, and the F1 indicator was improved. The experimental results showed that both the feature fusion module and the attention module had a positive effect on the model.

3.3. Compared with Other Methods

In order to further verify the effectiveness of the FA-SSD model in the detection of insulator umbrella disk shedding, under the condition of ensuring the same feature extraction network and hyperparameters, the method in this paper was compared with the commonly used target detection algorithm at this stage. The compared methods included Faster R-CNN [27], YOLOV3 [25], and RetinaNet [38], and the results are shown in Figures 9 and 10 and Table 3.

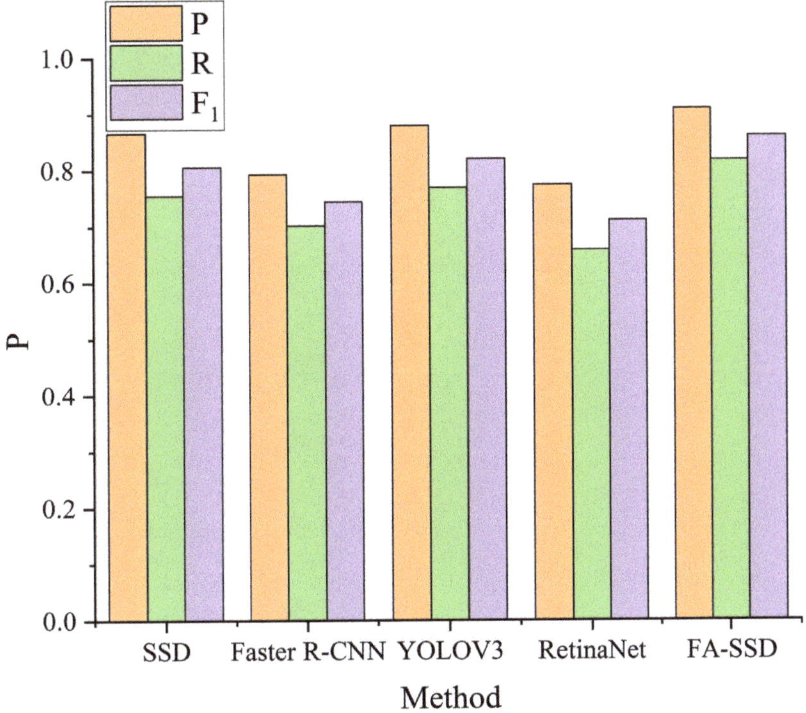

Figure 9. Results of different methods.

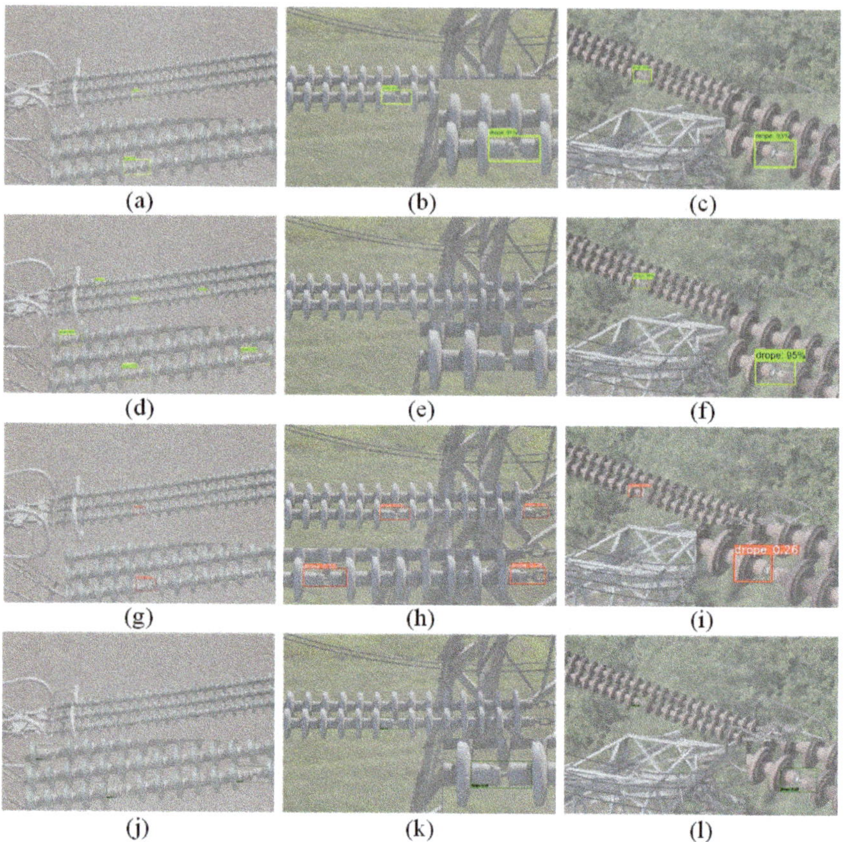

Figure 10. Visualization results of different methods. (**a–c**) Faster R-CNN. (**d–f**) YOLOV3. (**g–i**) RetinaNet. (**j–l**) FA-SSD.

Table 3. Results of different methods.

Method	Input Size	P	R	F_1
SSD [32]	300×300	0.866	0.755	0.806
Faster R-CNN [27]	800×800	0.793	0.702	0.744
YOLOV3 [25]	300×300	0.879	0.769	0.820
RetinaNet [38]	300×300	0.774	0.658	0.711
FA-SSD	300×300	**0.909**	**0.817**	**0.860**

It can be seen that FA-SSD significantly outperformed SSD and other commonly used object detection algorithms. Compared with other algorithms, the accuracy rate of detecting the umbrella disc shedding was improved on average 8.1%, and the recall rate was improved on average 9.6%. Compared with other target detection algorithms, the FA-SSD algorithm improved the detection accuracy and reduced the missed detection rate.

3.4. Dehazing Algorithm Experiment

As shown in Figure 11, after using the dehazing algorithm to dehaze the hazy images, the pictures became clearer.

Figure 11. Visualization of Dehazing Algorithms. (**a**) Foggy images. (**b**) Images after dehazing.

In order to verify the effectiveness of the dehazing algorithm proposed in this paper for the detection of insulator umbrella disc shedding in foggy images, the dehazing algorithm proposed in this paper was combined with the target detection algorithm, and the obtained detection results are shown in Figure 12.

As shown in Figure 12, the accuracy and recall of the model proposed in this paper were better than other models. It can be seen that after adding the defogging model, the accuracy and recall rate of the insulator umbrella disc shedding detection of the other models were significantly improved. Among them, the accuracy rate of the model increased by 0.08 on average, and the recall rate increased by 0.06 on average. This is because the clear image obtained by the dehazing algorithm was more conducive to the extraction of the features, thereby improving the detection effect. As shown in Figure 13, after adding the defogging algorithm, the detection effect of the FA-SSD model was significantly improved.

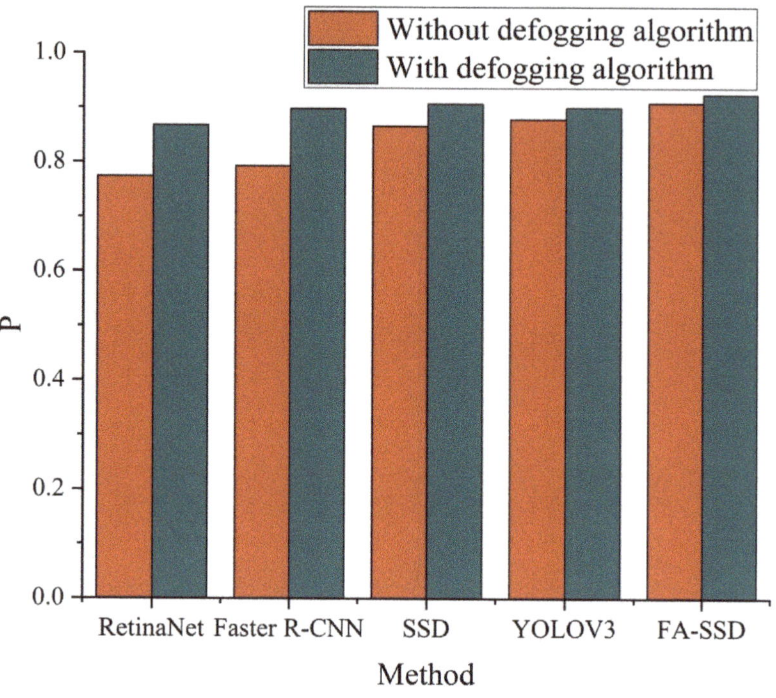

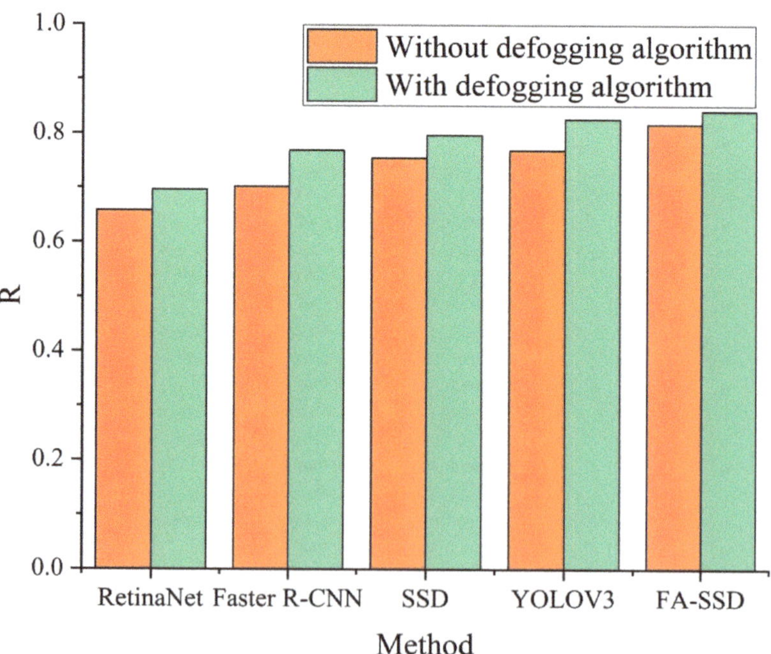

Figure 12. Experimental results before and after adding the defogging algorithm.

(a)   (b)

Figure 13. Visualization results of the FA-SSD and FA-SSD with defogging algorithm. (**a**) FA-SSD. (**b**) FA-SSD with defogging algorithm.

4. Discussion

On the basis of realizing the defect detection of insulators with foggy images, combined with the high-speed transmission advantages of 5G technology, real-time detection of insulator defects can be realized, and the necessary processing methods can be taken in time to reduce insulator failures. Compared with the traditional manual inspection, the method in this paper can reduce labor, material resources, and the influence of subjective factors; compared with the currently used UAV inspection, the method in this paper is more in real time. In the context of China's vigorous promotion of a smart grid, this research has important practical significance and good development prospects.

In the future, our research will have the following three aspects. First, we will examine more dehazing algorithms, such as the latest semi-supervised [39] or unsupervised [40] frameworks. Second, we will collect more fogged images of insulators and conduct a joint training strategy to combine image dehazing with defect detection [41]. Third, we will study the defect detection of insulators under a series of complex weather conditions such as sand, rain, and snow and devote ourselves to solving the problem of the defect detection of transmission lines in complex weather, so as to realize all-weather real-time monitoring of transmission lines.

5. Conclusions

Aiming to solve the difficulty of fully extracting effective features from foggy insulator images, as well as the small and difficult to detect proportion of umbrella disk shedding in an image, this paper proposed a detection method for insulator umbrella disk shedding defects that combined a dehazing algorithm and FA-SSD. Through the two-stage algorithm of dehazing and detection, the accurate detection of the insulator umbrella disk shedding in a foggy image was realized. This paper is the first to detect the defects in transmission lines with foggy images, which provides a solution for all-weather monitoring of transmission lines under complex weather conditions.

Author Contributions: Conceptualization, R.X. and X.C.; methodology, K.Y. and X.W.; software, J.W.; validation, R.X., X.C. and J.W.; formal analysis, K.Y.; investigation, X.W.; resources, Y.Z.; data curation, X.W.; writing—original draft preparation, K.Y.; writing—review and editing, R.X.; visualization, J.W.; supervision, X.C.; project administration, Y.Z.; funding acquisition, R.X. All authors have read and agreed to the published version of the manuscript.

Funding: This paper was supported by the Science and Technology Project of State Grid Hebei Information and tTelecommunication Branch: Research projects on key technologies and typical application scenarios based on "5G+AR+AI" (No.SGHEXT00SJJS2100036), and this research was funded by the National Natural Science Foundation of China Joint Fund Project Key Support Project: Robot semantic perception and interaction collaboration based on cross-spectral multisensory information fusion in visibility-limited environments (project number U21A20486).

Institutional Review Board Statement: Not applicable.

Informed Consent Statement: Not applicable.

Data Availability Statement: Not applicable.

Acknowledgments: Special thanks are given to the Information and Communication Branch, State Grid Hebei Electric Power and the Department of Automation and North China Electric Power University.

Conflicts of Interest: The authors declare no conflict of interest. The funders had no role in the design of the study; in the collection, analyses or interpretation of data; in the writing of the manuscript; or in the decision to publish the results.

Abbreviations

The following abbreviations are used in this manuscript:

5G	Fifth Generation Mobile Communication Technology
AI	Artificial Intelligence
HD	High Definition
SSD	Single Shot MultiBox Detector
NMS	Non-maximum suppression
YOLO	You Only Look Once
CNN	Convolutional Neural Network
FA-SSD	SSD Combining Feature Fusion and Attention Mechanism
FPN	Feature Pyramid Network
AR	Augmented Reality
GPU	Graphics Processing Unit
P	Precision
R	Recall
TP	True Positive
FP	False Positive
FN	False Negative
UAV	Unmanned Aerial Vehicle

References

1. Asprou, M.; Kyriakides, E.; Albu, M.M. Uncertainty bounds of transmission line parameters estimated from synchronized measurements. *IEEE Trans. Instrum. Meas.* **2018**, *68*, 2808–2818. [CrossRef]
2. Zhao, Z.; Qi, H.; Qi, Y.; Zhang, K.; Zhai, Y.; Zhao, W. Detection method based on automatic visual shape clustering for pin-missing defect in transmission lines. *IEEE Trans. Instrum. Meas.* **2020**, *69*, 6080–6091. [CrossRef]
3. Park, K.C.; Motai, Y.; Yoon, J.R. Acoustic fault detection technique for high-power insulators. *IEEE Trans. Ind. Electron.* **2017**, *64*, 9699–9708. [CrossRef]
4. Zhai, Y.; Wang, D.; Zhang, M.; Wang, J.; Guo, F. Fault detection of insulator based on saliency and adaptive morphology. *Multimed. Tools Appl.* **2017**, *76*, 12051–12064. [CrossRef]
5. Xia, H.; Yang, B.; Li, Y.; Wang, B. An Improved CenterNet Model for Insulator Defect Detection Using Aerial Imagery. *Sensors* **2022**, *22*, 2850. [CrossRef]
6. Wen, Q.; Luo, Z.; Chen, R.; Yang, Y.; Li, G. Deep learning approaches on defect detection in high resolution aerial images of insulators. *Sensors* **2021**, *21*, 1033. [CrossRef]
7. Deng, C.; Wang, S.; Huang, Z.; Tan, Z.; Liu, J. Unmanned Aerial Vehicles for Power Line Inspection: A Cooperative Way in Platforms and Communications. *J. Commun.* **2014**, *9*, 687–692. [CrossRef]
8. Zhai, Y.; Yang, X.; Wang, Q.; Zhao, Z.; Zhao, W. Hybrid knowledge r-cnn for transmission line multifitting detection. *IEEE Trans. Instrum. Meas.* **2021**, *70*, 1–12. [CrossRef]
9. Zhang, C.; Ueng, Y.L.; Studer, C.; Burg, A. Artificial intelligence for 5G and beyond 5G: Implementations, algorithms, and optimizations. *IEEE J. Emerg. Sel. Top. Circuits Syst.* **2020**, *10*, 149–163. [CrossRef]
10. Mahmood, A.; Beltramelli, L.; Abedin, S.F.; Zeb, S.; Mowla, N.; Hassan, S.A.; Sisinni, E.; Gidlund, M. Industrial IoT in 5G-and-beyond networks: Vision, architecture, and design trends. *IEEE Trans. Ind. Inform.* **2021**, *18*, 4122–4137. [CrossRef]
11. Liu, X.; Li, Y.; Shuang, F.; Gao, F.; Zhou, X.; Chen, X. Issd: Improved ssd for insulator and spacer online detection based on uav system. *Sensors* **2020**, *20*, 6961. [CrossRef] [PubMed]
12. Stark, J.A. Adaptive image contrast enhancement using generalizations of histogram equalization. *IEEE Trans. Image Process.* **2000**, *9*, 889–896. [CrossRef] [PubMed]
13. Liu, X.; Zhang, H.; Cheung, Y.m.; You, X.; Tang, Y.Y. Efficient single image dehazing and denoising: An efficient multi-scale correlated wavelet approach. *Comput. Vis. Image Underst.* **2017**, *162*, 23–33. [CrossRef]
14. Li, C.; Tang, S.; Kwan, H.K.; Yan, J.; Zhou, T. Color correction based on cfa and enhancement based on retinex with dense pixels for underwater images. *IEEE Access* **2020**, *8*, 155732–155741. [CrossRef]
15. He, K.; Sun, J.; Tang, X. Single image haze removal using dark channel prior. *IEEE Trans. Pattern Anal. Mach. Intell.* **2010**, *33*, 2341–2353.
16. Zhou, F.; Meng, X.; Feng, Y.; Su, Z. SNPD: Semi-Supervised Neural Process Dehazing Network with Asymmetry Pseudo Labels. *Symmetry* **2022**, *14*, 806. [CrossRef]
17. Chen, J.; Yang, G.; Ding, X.; Guo, Z.; Wang, S. Robust detection of dehazed image via dual-stream CNNs with adaptive feature fusion. *Comput. Vis. Image Underst.* **2022**, *217*, 103357. [CrossRef]
18. Zhao, W.; Zhao, Y.; Feng, L.; Tang, J. Attention Enhanced Serial Unet++ Network for Removing Unevenly Distributed Haze. *Electronics* **2021**, *10*, 2868. [CrossRef]
19. Zhou, Z.; Rahman Siddiquee, M.M.; Tajbakhsh, N.; Liang, J. Unet++: A nested u-net architecture for medical image segmentation. In *Deep Learning in Medical Image Analysis and Multimodal Learning for Clinical Decision Support*; Springer: Berlin/Heidelberg, Germany, 2018; pp. 3–11.
20. Gao, G.; Cao, J.; Bao, C.; Hao, Q.; Ma, A.; Li, G. A Novel Transformer-Based Attention Network for Image Dehazing. *Sensors* **2022**, *22*, 3428. [CrossRef]
21. Zhang, Z.; Huang, S.; Li, Y.; Li, H.; Hao, H. Image Detection of Insulator Defects Based on Morphological Processing and Deep Learning. *Energies* **2022**, *15*, 2465. [CrossRef]
22. Tao, X.; Zhang, D.; Wang, Z.; Liu, X.; Zhang, H.; Xu, D. Detection of power line insulator defects using aerial images analyzed with convolutional neural networks. *IEEE Trans. Syst. Man Cybern. Syst.* **2018**, *50*, 1486–1498. [CrossRef]
23. She, L.; Fan, Y.; Wang, J.; Cai, L.; Xue, J.; Xu, M. Insulator Surface Breakage Recognition Based on Multiscale Residual Neural Network. *IEEE Trans. Instrum. Meas.* **2021**, *70*, 1–9. [CrossRef]
24. Zhang, X.; Zhang, Y.; Liu, J.; Zhang, C.; Xue, X.; Zhang, H.; Zhang, W. InsuDet: A Fault Detection Method for Insulators of Overhead Transmission Lines Using Convolutional Neural Networks. *IEEE Trans. Instrum. Meas.* **2021**, *70*, 1–12. [CrossRef]
25. Redmon, J.; Farhadi, A. Yolov3: An incremental improvement. *arXiv* **2018**, arXiv:1804.02767.
26. Zhao, W.; Xu, M.; Cheng, X.; Zhao, Z. An Insulator in Transmission Lines Recognition and Fault Detection Model Based on Improved Faster RCNN. *IEEE Trans. Instrum. Meas.* **2021**, *70*, 1–8. [CrossRef]
27. Ren, S.; He, K.; Girshick, R.; Sun, J. Faster r-cnn: Towards real-time object detection with region proposal networks. *Adv. Neural Inf. Process. Syst.* **2015**, *28*. Available online: https://proceedings.neurips.cc/paper/2015/hash/14bfa6bb14875e45bba028a21ed38046-Abstract.html (accessed on 17 May 2022). [CrossRef]
28. Lin, T.Y.; Dollár, P.; Girshick, R.; He, K.; Hariharan, B.; Belongie, S. Feature pyramid networks for object detection. *Proc. IEEE Conf. Comput. Vis. Pattern Recognit.* **2017**, *36*, 2117–2125.

29. Liu, W.; Ren, G.; Yu, R.; Guo, S.; Zhu, J.; Zhang, L. Image-Adaptive YOLO for Object Detection in Adverse Weather Conditions. *arXiv* **2021**, arXiv:2112.08088.

30. He, Y.; Liu, Z. A Feature Fusion Method to Improve the Driving Obstacle Detection Under Foggy Weather. *IEEE Trans. Transp. Electrif.* **2021**, *7*, 2505–2515. [CrossRef]

31. Hassaballah, M.; Kenk, M.A.; Muhammad, K.; Minaee, S. Vehicle detection and tracking in adverse weather using a deep learning framework. *IEEE Trans. Intell. Transp. Syst.* **2020**, *22*, 4230–4242. [CrossRef]

32. Liu, W.; Anguelov, D.; Erhan, D.; Szegedy, C.; Reed, S.; Fu, C.Y.; Berg, A.C. Ssd: Single shot multibox detector. In Proceedings of the European Conference on Computer Vision, Amsterdam, The Netherlands, 8–16 August 2016; Springer: Berlin/Heidelberg, Germany, 2016; pp. 21–37.

33. Chen, Z.; Wang, Y.; Yang, Y.; Liu, D. PSD: Principled synthetic-to-real dehazing guided by physical priors. In Proceedings of the Proceedings of the IEEE/CVF Conference on Computer Vision and Pattern Recognition, Nashville, TN, USA, 20–25 June 2021; pp. 7180–7189.

34. Li, B.; Ren, W.; Fu, D.; Tao, D.; Feng, D.; Zeng, W.; Wang, Z. Benchmarking single-image dehazing and beyond. *IEEE Trans. Image Process.* **2018**, *28*, 492–505. [CrossRef] [PubMed]

35. Li, Z.; Zhou, F. FSSD: Feature fusion single shot multibox detector. *arXiv* **2017**, arXiv:1712.00960.

36. He, K.; Zhang, X.; Ren, S.; Sun, J. Deep residual learning for image recognition. In Proceedings of the IEEE Conference on Computer Vision and Pattern Recognition, Las Vegas, NV, USA, 27–30 June 2016; pp. 770–778.

37. Hu, J.; Shen, L.; Sun, G. Squeeze-and-excitation networks. In Proceedings of the IEEE Conference on Computer Vision and Pattern Recognition, Salt Lake City, UT, USA, 18–23 June 2018; pp. 7132–7141.

38. Lin, T.Y.; Goyal, P.; Girshick, R.; He, K.; Dollár, P. Focal loss for dense object detection. In Proceedings of the IEEE International Conference on Computer Vision, Venice, Italy, 22–29 October 2017; pp. 2980–2988.

39. Li, Y.; Chang, Y.; Gao, Y.; Yu, C.; Yan, L. Physically Disentangled Intra-and Inter-Domain Adaptation for Varicolored Haze Removal. In Proceedings of the IEEE/CVF Conference on Computer Vision and Pattern Recognition, New Orleans, LA, USA, 21–24 June 2022; pp. 5841–5850.

40. Yuntong, Y.; Changfeng, Y.; Yi, C.; Lin, Z.; Xile, Z.; Luxin, Y.; Yonghong, T. Unsupervised Deraining: Where Contrastive Learning Meets Self-similarity. *arXiv* **2022**, arXiv:2203.11509.

41. Li, Y.; Chang, Y.; Yu, C.; Yan, L. Close the Loop: A Unified Bottom-Up and Top-Down Paradigm for Joint Image Deraining and Segmentation. 2022. Available online: https://www.aaai.org/AAAI22Papers/AAAI-678.LiY.pdf (accessed on 17 May 2022).

Article

Two-Level Model for Detecting Substation Defects from Infrared Images

Bing Li, Tian Wang, Zhedong Hu, Chao Yuan * and Yongjie Zhai

Department of Automation, North China Electric Power University, Baoding 071003, China
* Correspondence: chaoyuan@ncepu.edu.cn; Tel.: +86-134-0036-7541

Abstract: Training a deep convolutional neural network (DCNN) to detect defects in substation equipment often requires many defect datasets. However, this dataset is not easily acquired, and the complex background of the infrared images makes defect detection even more difficult. To alleviate this issue, this article presents a two-level defect detection model (TDDM). First, to extract the target equipment in the image, an instance segmentation module is constructed by training from the instance segmentation dataset. Then, the target equipment is segmented by the superpixel segmentation algorithm into superpixels according to obtain more details information. Next, a temperature probability density distribution is constructed with the superpixels, and the defect determination strategy is used to recognize the defect. Finally, experiments verify the effectiveness of the TDDM according to the defect detection dataset.

Keywords: infrared image; substation equipment; defect detection; superpixel segmentation; temperature probability density

1. Introduction

Substation equipment is an essential part of the power system [1]. Once defects exist in operating equipment, an abnormal temperature usually occurs at the defective parts, triggering thermal failures that can lead to local equipment burnout or even more severe electric power accidents [2]. Therefore, timely and accurate detection of defects in substation equipment is of great significance to the safety and stability of a power system.

Many methods have been studied for defects detection in substation equipment, including dielectric loss measurement [3], UHF (ultra-high frequency) method [4], FDR (frequency domain reflectometry) method [5], and infrared image-based methods [6,7]. The dielectric loss measurement requires off-line preventive testing, which will delay the operation of substation equipment. The complexity of the UHF method makes directly locating defective regions difficult. The FDR method is sensitive only to defects caused by moisture. Early infrared image-based methods for detecting thermal defects in substation equipment require manual intervention, which is time-consuming and costly. However, with the development of smart grids and the successful application of substation inspection robots, a large number of on-site infrared images needed to be inspected urgently. Intelligent defect detection methods have emerged based on computer vision.

Due to the redundant background and the densely packed targets, applying automatic intelligent defect detection methods directly is difficult. Thus, extracting the target equipment in the complex infrared images is required first. Early researchers studied the methods using traditional digital image processing techniques, including threshold-based, region-based, and edge-based methods. Threshold-based methods separate the foreground from the background of an image by selecting a suitable threshold [8], which is simple and efficient but susceptible to noise interference, causing poor robustness. A typical region-based method is the watershed algorithm [9]. It uses the local minima of the image gradient to form a specific region to segment different image parts. However, it

Citation: Li, B.; Wang, T.; Hu, Z.; Yuan, C.; Zhai, Y. Two-Level Model for Detecting Substation Defects from Infrared Images. *Sensors* **2022**, *22*, 6861. https://doi.org/10.3390/s22186861

Academic Editor: Jiayi Ma

Received: 18 August 2022
Accepted: 6 September 2022
Published: 10 September 2022

is sensitive to the color changes in the object's surface, giving rise to over-segmentation. Edge-based methods extract edge features from the image by edge detection operators such as the Sobel operator [10] and Canny operator [11] to realize the segmentation of an image. Nevertheless, it cannot guarantee the existence of closed, continuous edge regions, and it lacks robustness to noise interference. The recent rapid development of deep learning and imaging technologies has brought innovative ideas for extracted methods from infrared images of substation equipment. Instance segmentation is a classic task in the field of computer vision, which can perform object extraction excellently in images. This task, not only locates and classifies all instances but also segments each instance from the images [12]. Many applications benefit from accurate instance segmentation, including electrical systems [13,14], autonomous driving [15], robotics [16,17], and intelligent transportation systems [18]. Consequently, instance segmentation has become an active research topic in the industry, which benefits its powerful ability of object extraction. Xiong et al. [19] proposed a method based on Mask R-CNN and Bayesian context network to recognize power equipment, which is considered the relationship between objects in a complex background. Ling et al. [20] presented a novel deep learning framework to locate the broken insulators, which is address the problem of low signal-noise-ratio (SNR) setting. To detect the transmission line, a transmission line detection (TLD) algorithm is proposed [21], which is a multitask deep neural network with branched outputs. The deep learning-based methods show excellent performance to extract the target object.

In the stage of defect detection, some promising methods for detecting defects are feature extraction and convolutional neural networks. The key to feature extraction-based approaches is acquiring target ontology features and using classifiers to recognize the extracted features [22,23]. However, the effectiveness of feature extraction and the selection of classifiers are great dependence on personal experience. Convolutional neural networks focus on detecting target defects through an object detection model [24,25]. Li et al. [26] proposed a method of insulator defect location, which is cascades detection and segmentation networks from two levels. In view of the characteristics of insulator defects, Wang et al. [27] presented an improved network to detect the defect of aerial insulator photos. The above method achieved excellent results in defect detection, but requires numerous defective insulator images to train the DCNN. In reality, the infrared images of defective substation equipment are difficult to acquire, and the performance of DCNN is difficult to guarantee. Implementing defect detection of substation equipment in infrared images is still challenging. In an infrared image, the different parts of the target corresponding to different heat generation characteristics. Thus, the temperature feature of the target is used to estimate temperature probability density distribution, which is used to identify defects by the presented strategy. The proposed defect detection part is an unsupervised learning method and is not limited by the dataset. Before that, the superpixel processing is used to provide more details, those details offer more information for defect detection. Meanwhile, it reduces the complexity and time spent on the model.

This study proposes the TDDM for defect detection in electric power substations, which is used in infrared images of substation equipment, e.g., insulator, current transformer, lightning arrester, bushing and voltage transformer. The main contributions of this paper are as follows.

(1) Inspecting the substation equipment from the infrared images with the redundant background and the densely packed targets directly is difficult. The proposed TDDM extracted the target firstly, and then, defect analysis is conducted on a single instance, which is converted to a two-level detection problem.

(2) Superpixel segmentation is conducted on the extracted target equipment to merge adjacent pixels with similar characteristics. The process is used to provide more details and reduce the complexity of the subsequent detection determination.

(3) Based on a Gaussian kernel function, the temperature probability density distribution of the target equipment is constructed, which is used in a defect determination strategy to find the defective areas in infrared images of the target substation equipment.

(4)   The experimental results show that the proposed model accurately detects defects in substation equipment in infrared images.

The remainder of this paper is organized as follows. In Section 2, a novel model for detecting these defects in infrared images is provided, including instance segmentation, superpixel segmentation, and defect determination. Section 3 verifies the performance of the proposed model and discusses the influences of superpixel parameters on the results. Section 4 concludes this work.

2. Procedure for the Proposed Model

The model proposed is designed for automatically detecting defects of substation equipment in infrared images. The model transforms defect detection into a two-level detection problem. First, an instance segmentation algorithm directly extracts the target equipment from infrared images with complex backgrounds. After that, a superpixel segmentation algorithm merges similar characteristics and captures the details of the target equipment. Finally, the defect position is determined. Figure 1 is a flowchart of the proposed TDDM procedure.

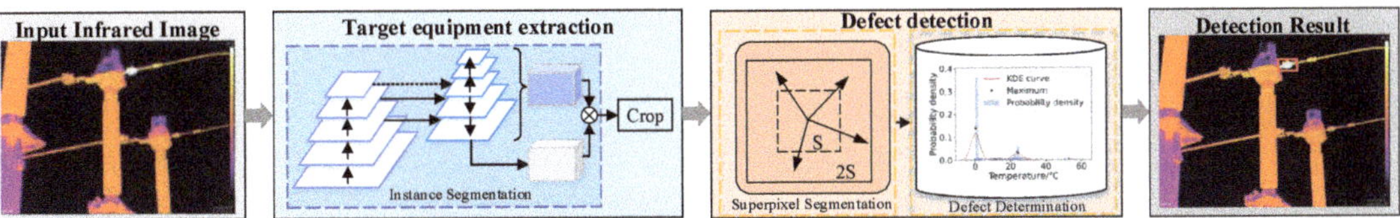

Figure 1. Flowchart of TDDM.

2.1. Instance Segmentation

To detect substation equipment in infrared images, we must first extract the target equipment from the image. Instance segmentation is a basic task of DCNN, which is to extract the target from a complex background and distinguish different instances in the image's foreground [28]. There are three commonly used instance segmentation methods: top-down detection-based methods, bottom-up semantic segmentation-based methods and direct instance segmentation at the pixel level. Top-down detection-based methods perform instance segmentation in a bounding box, such as the Mask R-CNN [29], Mask Scoring R-CNN [30], and YOLACT [31] networks. In bottom-up semantic segmentation-based methods, the pixels are labeled for prediction and clustered [32,33]. The SOLO algorithm [34] performs end-to-end optimization of instance segmentation by mask labeling, which directly segments instances at the pixel level.

This study extracted target equipment images using YOLACT. Its backbone is the feature extraction part used to obtain different resolution feature maps $C_i (i = 2, 3, 4, 5)$ from the input infrared image. The description of specific backbone configuration parameters as shown in Table 1. To obtain the multiscale features, $C_i (i = 2, 3, 4, 5)$ are fused by the horizontal connection with the feature pyramid. Then multiscale features $P_j (j = 3, 4, 5, 6, 7)$ are connected to prediction heads for multiscale prediction of objects. There are two branches after the feature pyramid. The one branch predicts the object category, the bounding box, and the mask coefficients; the higher score bounding box is obtained through non-maximum suppression (NMS) [35]. The other branch is a fully convolutional network called protonet, which generates a series of prototype masks based on the feature map P_3. Finally, the prototype masks obtained from protonet are linearly combined with mask coefficients to get m instance $c_m (m \in \{1, 2, \cdots, M\}$. We can perform defect analysis on a single instance, removing interference from complex backgrounds.

Table 1. The description of specific backbone configuration parameters.

Layer Name	Structure	Convolution Kernel	Feature Map Size
Input Layer	-	-	640 × 640
Conv1	-	7 × 7 × 64, stride 2	320 × 320
Pool1	Maxpool	3 × 3 × 64, stride 2	160 × 160
Conv2_x	Bottleneck	$\begin{bmatrix} 1 \times 1 \times 64 \\ 3 \times 3 \times 64 \\ 1 \times 1 \times 256 \end{bmatrix} \times 3$	160 × 160
Conv3_x	Bottleneck	$\begin{bmatrix} 1 \times 1 \times 128 \\ 3 \times 3 \times 128 \\ 1 \times 1 \times 512 \end{bmatrix} \times 4$	80 × 80
Conv4_x	Bottleneck	$\begin{bmatrix} 1 \times 1 \times 256 \\ 3 \times 3 \times 256 \\ 1 \times 1 \times 1024 \end{bmatrix} \times 6$	40 × 40
Conv5_x	Bottleneck	$\begin{bmatrix} 1 \times 1 \times 512 \\ 3 \times 3 \times 512 \\ 1 \times 1 \times 2048 \end{bmatrix} \times 3$	20 × 20

2.2. Superpixel Segmentation

In the previous section, the image of each type of target equipment in the infrared image is segmented. In this section, the target equipment is detected individually. To make defect detection easier, we first perform superpixel segmentation. Superpixel segmentation forms superpixels from adjacent pixels in the image of target equipment with similar texture, color, luminance, or other characteristics. Thus, superpixels can be treated as processing units, reducing the complexity and time spent on the subsequent processing of the image [36]. Superpixel segmentation methods are generally classified into graph theory-based methods [37,38] and clustering-based methods [39–41]. Computation of cost functions in graph theory-based methods is complicated. In contrast, clustering-based methods has simple principles and good interpretability. The clustering-based simple linear iterative clustering (SLIC) algorithm obtains uniform compact superpixels, and it has good controllability and low operational complexity than other superpixel algorithms [42].

Inspired by the SLIC algorithm, the proposed model forms adjacent pixels with similar temperature characteristics t and spatial characteristics into superpixels $c_m^n, n \in \{1, 2, \ldots, N\}$. Assume that there are I pixels in infrared image c, and the number of superpixels is K. Then the interval between the clustering centers C_k is $S = \sqrt{I/K}$. The pixels $2S$ distance from the clustering center is iteratively clustered based on spatial similarity and temperature similarity, until the maximum number of iterations is reached. The formula for calculating the distance D between pixel i and the cluster center C_k is as follows:

$$D = \sqrt{\left(\frac{d_t}{m_t}\right)^2 + \left(\frac{d_{xy}}{m_{xy}}\right)^2},$$

(1)

$$d_t = \sqrt{(t_k - t_i)^2},$$

(2)

$$d_{xy} = \sqrt{(x_k - x_i)^2 + (y_k - y_i)^2},$$

(3)

where d_t is the temperature distance between pixel i and the cluster center C_k, d_{xy} is the spatial distance between pixel i and the cluster center C_k, m_t and m_{xy} are the maximum temperature distance and spatial distance obtained in the previous iteration, respectively.

Further, the superpixels c_m^n of each instance are obtained, and the corresponding temperature characteristic $T_m^n, n \in \{1, 2, \ldots, N\}$ is calculated by averaging the temperature of pixels in the superpixel. All temperature characteristics of c_m^n lie between the maximum temperature $T_m^{\max}$ and the minimum temperature $T_m^{\min}$, i.e., $T_m^n \in [T_m^{\min}, T_m^{\max}]$.

2.3. Defect Determination

After superpixel segmentation of the target equipment, we inspect the target equipment one by one to determine whether there exist defects. Figure 2 shows the target equipment of the background, normal region, and defective region with different temperature characteristics in the infrared image. The range of temperatures that the defect determination algorithm can identify is even broader than the temperatures range in Figure 2.

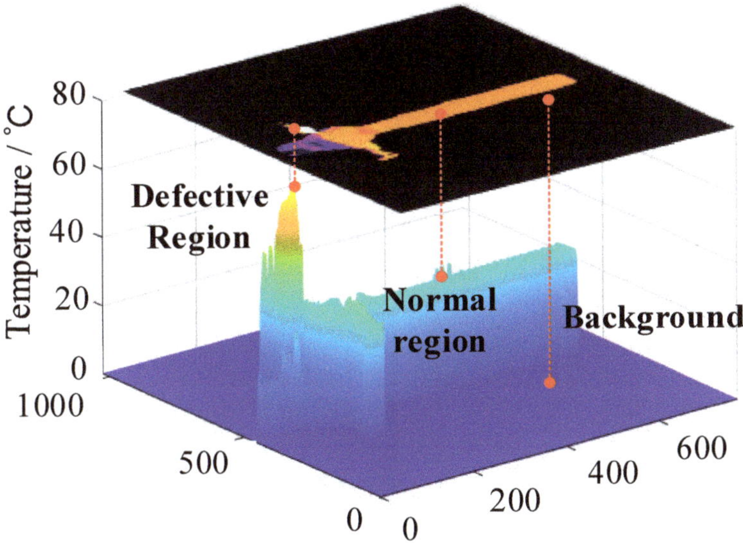

Figure 2. Infrared image of substation equipment and its temperature distribution.

Different temperature characteristics correspond to different temperature probability densities. Thus, we can model the temperature probability density distribution of the instances to determine whether there are defects.

For instance c_m, the temperature probability density T_m^n can be calculated by Equation (4), as shown by the blue histogram in Figure 3.

$$f_m(n) = \frac{T_m^n}{\sum\limits_{i=1}^{N} T_m^i}, n \in [1, N]. \tag{4}$$

However, the temperature probability density data are discretized, which cannot be used directly. Thus, we need to estimate the probability density function to approximate its specific distribution. The common probability density estimation methods include parametric probability density estimation and non-parameter probability density estimation. Kernel density estimation (KDE) [43] is a non-parameter probability density estimation method used to estimate the temperature probability density distribution of the data.

If there is a sufficiently small temperature region $A = [T_m^{A\,\min}, T_m^{A\,\max}]$ with bandwidth h, the probability $P_m(A)$ of T_m^n in A is

$$P_m(A) = \int_A f_m(x)dx \approx f_m(x) \int_{T_m^{A\,\min}}^{T_m^{A\,\max}} dx = f_m(x)h. \tag{5}$$

Suppose the probability of Z out of N data falling into region A is

$$P_m(A) = \frac{Z}{N}. \tag{6}$$

Then the temperature probability density becomes

$$f_m(x) = \frac{Z}{Nh}. \tag{7}$$

The kernel density estimation of Equation 7 using the Gaussian kernel function obtains the temperature probability density function of instance c_m.

$$f_m(x) = \frac{1}{Nh}\sum_{j=1}^{N}\frac{1}{\sqrt{2\pi}}e^{-\frac{1}{2}\left(\frac{x-T_m^j}{h}\right)^2}. \tag{8}$$

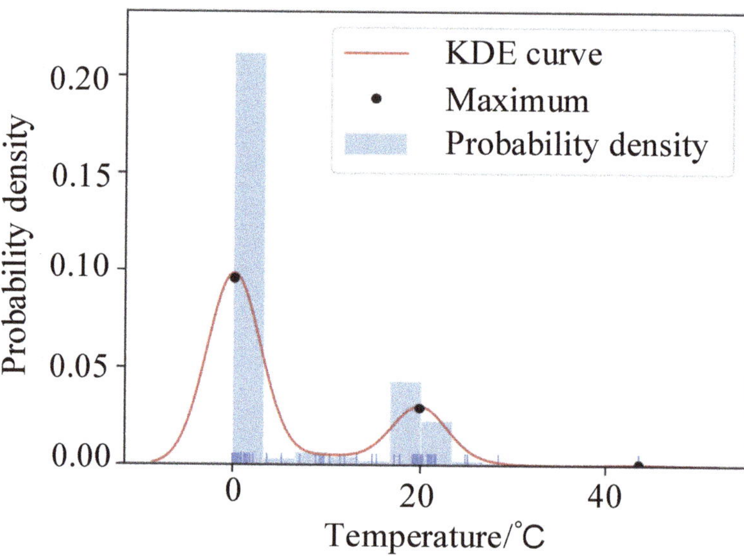

Figure 3. Temperature probability density distribution of c_m.

After that, the temperature probability density distribution function is visualized in Figure 3 by the red curve. The point of local maximum $O_m^q((x_m^q, y_m^q)), q = 1, 2, \ldots, Q)$ is obtained, which is denoted by the black dots in Figure 3.

Based on the temperature probability density distribution function, we propose a determination strategy to find defects in infrared images. Due to the different temperature characteristics in the background, normal region, and defective region. Meanwhile, different temperature areas are shown in the temperature probability density distribution. Thus, the presence of O_m^q and $Q \geq 3$ indicate the presence of a defect in the target equipment in this strategy. Then, through the application of the proposed algorithm, x_m^3 is used as the threshold, superpixels c_m^n with temperature characteristics T_m^n higher than x_m^3 are determined to be defective superpixels, automatically. Then, all adjacent defective superpixels are merged to determine the defective regions D_m in instance c_m. Finally, all instances of the infrared image are traversed to obtain all the defective regions automatically. In addition, Algorithm 1 summarizes the whole programming procedure of the proposed TDDM.

Algorithm 1 TDDM

1: **Input:**Infrared image c, Number of superpixels K.
2: **Output:** All defect regions in the infrared image.
3: Obtain instance $c_m = \text{Seg}(c), m = 1, 2, \ldots, M$
4: **for** $m = 1 \text{to} M$ **do**
5:    **for** $n = 1 \text{to} N$ **do**
6:       Compute superpixels c_m^n
7:       Obtain temperature characteristic T_m^n
8:    **end for**
9:    Compute temperature probability density distribution f_m
10:    Compute the local maximum $O_m^q(x_m^q, y_m^q)$ of f_m, where $q = 1, 2, \ldots, Q$
11:    **if** $Q \geq 3$ **then**
12:       **for** $n = 1 \text{to} N$ **do**
13:          **if** $T_m^n > x_m^3$ **then**
14:             Determine c_m^n defective
15:          **else**
16:             Determine c_m^n normal
17:          **end if**
18:       **end for**
19:       Merge all adjacent defective superpixels to obtain D_m
20:    **else**
21:       **Output:** No defect in the instance.
22:    **end if**
23: **end for**

---

3. Experiments

3.1. Data Preparation

The experimental infrared images in this article consist of five types of substation equipment, including insulator, current transformer, voltage transformer, bushing, and lightning arrester. The images were captured in a substation by the FLIR T600, where the infrared image resolution is 480×360 and the temperature resolution is 0.04 °C. The dataset composition of the substation equipment infrared images in the experiments is illustrated in Figure 4. The instance segmentation dataset is used to train the instance segmentation module, in which the dataset all consists of the normal substation equipment images. The number of each type of equipment is shown in Table 2. In addition, the defect detection dataset is used to evaluate the performance of the TDDM.

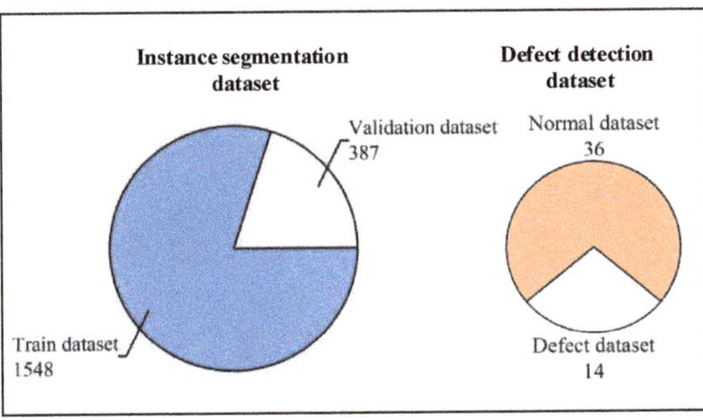

Figure 4. Dataset composition of the substation equipment infrared images.

Table 2. Number of each type of equipment in the instance segmentation dataset.

Equipment	Number
Insulator	919
Current transformer	413
Lightning arrester	289
Bushing	161
Voltage transformer	153

3.2. Instance Segmentation Results and Analysis

The instance segmentation algorithm ran on Ubuntu 18.04LTS with NVIDIA 2080Ti. The training was conducted under the network framework PyTorch through Python3.8, accelerated by CUDA11.2. The current advanced instance segmentation algorithms, including SOLO, Mask R-CNN, MS R-CNN, and YOLACT, were compared experimentally. For training the instance segmentation algorithm, the batch size was set to 2, the SGD optimizer was used, the momentum value was 0.9, the initial learning rate was 0.001, and the number of training iterations was 60 epochs.

To choose the optimal instance segmentation algorithm, a multi-target scene with a complex background was selected for testing. The performance indexes were mAP (mean average precision) and mAR (mean average recall), which are commonly used indexes in the current instance segmentation. SOLO, Mask R-CNN, Mask Scoring R-CNN, and YOLACT were tested on the instance segmentation dataset. The experiment results are shown in Figure 5 and Table 3.

In Table 3, YOLACT had the highest segmentation accuracies compared with the other three algorithms. The values are 67.0% and 74.0% in terms of the mAP and mAR metrics, which were 10.1% and 12.5% higher than the SOLO algorithms. As shown in Figure 5, Figure 5a are the original images and Figure 5f are the ground truth. The four algorithms are intuitively compared in Figure 5b–e, where the white rectangle represents the location of the substation equipment by the model. The pixels of instances belonging to the different categories are marked with different colors. It can be seen from Figure 5 that the YOLACT algorithm accurately located the substation equipment in infrared images and had typically higher quality masks. Thus, this study chose the YOLACT algorithm to segment substation equipment infrared images.

Table 3. Comparison of instance segmentation algorithms.

Method	mAP/%	mAR/%
SOLO	56.9	61.5
Mask R-CNN	63.6	70.4
Mask Scoring R-CNN	65.1	70.9
YOLACT	67.0	74.0

3.3. Compared with Other Superpixel Segmentation Methods

In this section, we compare SLIC [40] to several popular superpixel segmentation algorithms including Felzenszwalb [44], Quickshif [45], and Watershed [46] by the defect detection dataset. The performance of superpixel segmentation is quantitatively evaluated by two metrics, which are boundary recall (BR) and under-segmentation error (UE). BR is the most commonly used metric, which is the percentage of superpixels boundaries coinciding with ground truth boundaries.

$$BR = \frac{SP}{GP}, \tag{9}$$

where SP is the number of segmentation results that meet the condition that should be the ground truth. GP is the total number of the segmentation result. The higher the BR denotes the better performance. UE is the ratio of calculated over-segmented superpixels.

The more approaches zero of the UE, the superpixel approaches the ground truth. UE is defined as follow

$$UE = -1 + \frac{1}{N} \sum |u_m \cap u_n| > \omega |u_m| \, |u_n|, \tag{10}$$

where u_m and u_n are the pixel sets of the m-th superpixel and ground truth, respectively. ω is set to 0.05 for well established [47]. The lower the UE denotes the better performance.

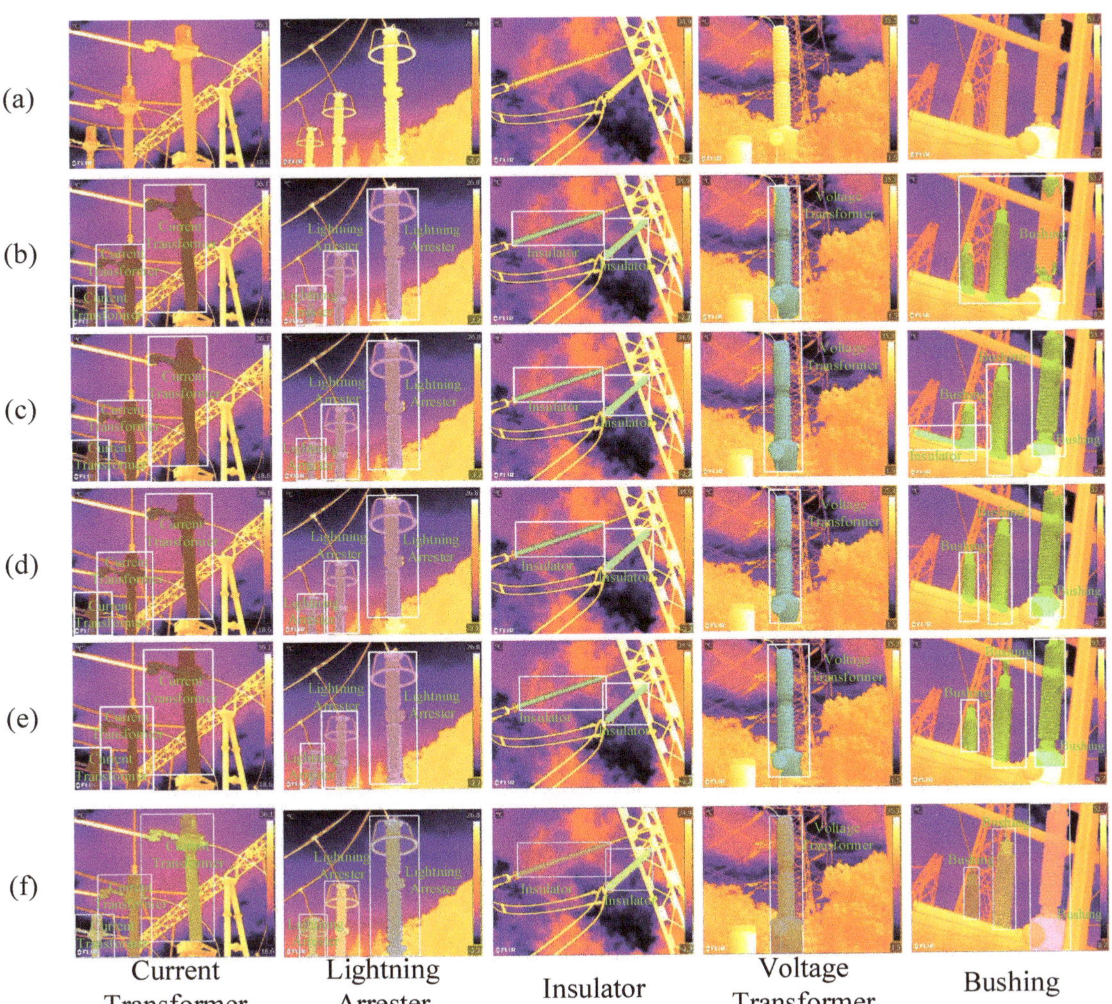

Figure 5. Comparison of segmentation results of different instance segmentation algorithms. (**a**) Original Images. (**b**) SOLO. (**c**) Mask R-CNN. (**d**) Mask Scoring R-CNN. (**e**) YOLACT. (**f**) Ground Truth.

As shown in Figure 6, it illustrates the comparative performance the methods on the defect detection dataset. The numbers of superpixels are set to 250, 500, 750, 1000, 1250, and 1500, respectively. From Figure 6, SLIC, Watershed, and Quickshif all obtain good performance since BR is higher than 0.86. The value of UE in SLIC is the lowest among all methods, this means that better compactness of superpixel segmentation can be achieved.

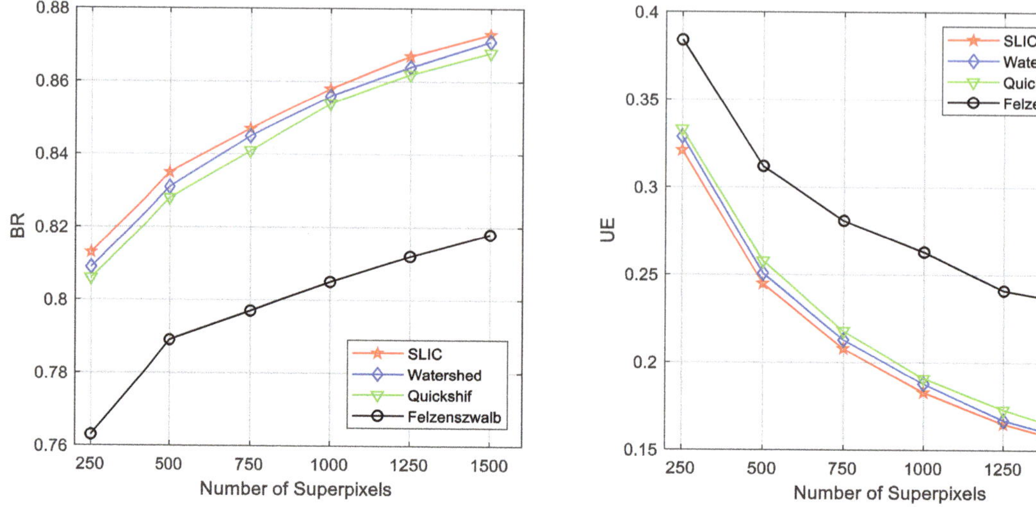

Figure 6. Comparison of superpixel segmentation algorithms performances.

3.4. Defect Detection Results and Analysis

We adopted four widely used metrics for the quantitative evaluations of defect detection performance: precision (P_r), recall (R_e), F_1, and mean running time (mRN). A higher evaluation value indicates better performance, calculated as follows.

$$P_r = \frac{\text{TP}}{\text{TP} + \text{FP}},\qquad(11)$$

$$R_e = \frac{\text{TP}}{\text{TP} + \text{FN}},\qquad(12)$$

$$F_1 = \frac{2 * P_r * R_e}{P_r + R_e},\qquad(13)$$

where TP and denote the number of correctly detected defects. TP + FP and TP + FN denote the total number of detected defects and the total number of actual defects, respectively. F_1 is the harmonic mean of P_r and R_e.

We use mean intersection over union (mIoU) to calculate the accuracy of defect region localization. The mIoU is defined as

$$\text{mIoU}(G_T, P_M) = \sum_{m=1}^{M} \frac{\text{Area}(G_T^m \cap P_D^m)}{\text{Area}(G_T^m \cup P_D^m)},\qquad(14)$$

where G_T^m is the ground truth and P_D^m is the predicted region.

To verify the effectiveness of the TDDM, the defect detection datasets are input to the TDDM. To choose the best parameter of the number of superpixels K, we set K from 250 to 3000 with an interval of 250 for the ablation experiments. When $K = 1000$, TDDM has achieved the best defect detection performance. The values of precision, recall, F_1, and mIoU were 0.812, 0.928, 0.866 and 0.831. When $K = 2250$, the model had acceptable precision and recall values performance, but the model running time became longer. Moreover, the running time of TDDM increased with K. Thus, in a word, the selection of an appropriate K is important. Table 4 and Figure 7 show the comparison with a different number of superpixels K to the defect detection dataset.

Table 4. Detection performance for different numbers of K.

Number of K	P_r	R_e	F_1	mIoU	mRN
250	0.315	0.428	0.362	0.145	1.47
500	0.190	0.285	0.228	0.098	1.91
750	0.555	0.714	0.624	0.582	2.73
1000	0.812	0.928	0.866	0.831	4.30
1250	0.705	0.857	0.773	0.683	4.36
1500	0.722	0.928	0.812	0.738	5.54
1750	0.764	0.928	0.838	0.796	5.64
2000	0.764	0.928	0.838	0.796	6.82
2250	0.812	0.928	0.866	0.817	6.68
2500	0.764	0.928	0.838	0.796	8.64
2750	0.764	0.928	0.838	0.796	10.42
3000	0.764	0.928	0.838	0.796	11.97

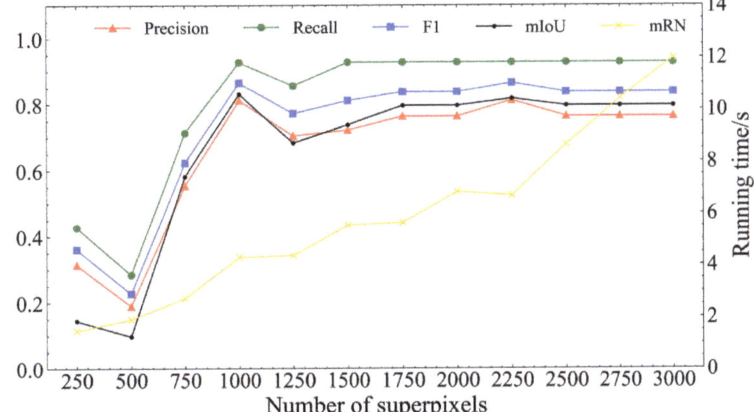

Figure 7. Results of ablation experiments on the number of superpixels.

To evaluate the superiority of the proposed method, some ablation experiments were performed on TDDM. (1) Evaluate the advantage of the superpixel segmentation algorithm (SSA) as a preprocessing for defect detection. (2) Evaluate the advantage of the DCNN + superpixel method for defect detection. Table 5 lists the results of the ablation experiment. As shown in Table 5, the SSA can provide more details and reduce the complexity of the subsequent detection determination. When the objects are extracted firstly by DCNN, the metrics for evaluating accuracy have improved. It indicates that DCNN can overcome the problem of complex background in infrared images. The model achieved superior results when both DCNN and SSA were used. P_r, R_e, F_1, mIoU are reached 0.812, 0.928, 0.866, and 0.831, respectively, which were the highest values.

Table 5. Ablation experiment of TDDM.

DCNN	SSA	P_r	R_e	F_1	mIoU	mRN
✓		0.764	0.928	0.838	0.796	21.34
	✓	0.555	0.714	0.624	0.582	3.62
✓	✓	0.812	0.928	0.866	0.831	4.30

As shown in Figures 8 and 9, the intuitive defect detection process of the TDDM in this paper is on the defect detection dataset. In the intuitive experiment results, the different categories have displayed.

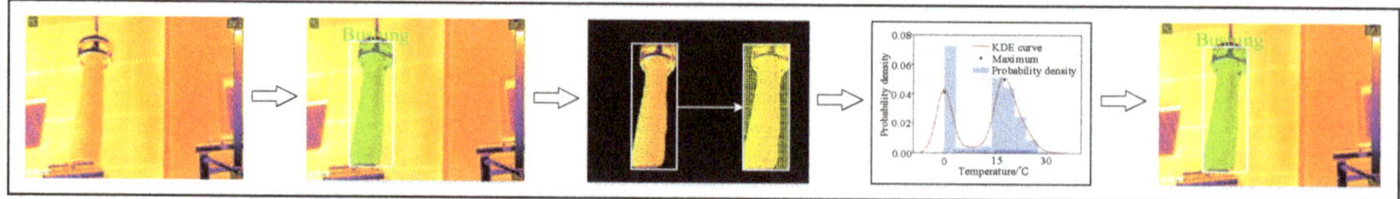

Figure 8. Process of the normal bushing infrared image detection.

Figure 8 shows the process of the normal bushing infrared image detection. In the fourth column, the temperature probability density distribution of the bushing has only two local maxima, which reflects that the substation equipment is no defect. This demonstrates that the TDDM is effectively applied in detecting normal substation equipment.

Figure 9 shows the entire detection flow of the TDDM to the defect-located infrared images. From left to right are the input infrared images, instance segmentation, superpixel segmentation, defect determination, and defect detection results. At the penultimate column, there are three maxima in the temperature probability density distribution of target equipment, representing the equipment exist defect. The target equipment defect detection results are shown in the last column. The white rectangle denotes the target equipment, and the red rectangle represents the location of the defective regions. As can be seen that, TDDM accurately located the defect in substation equipment against a complex background.

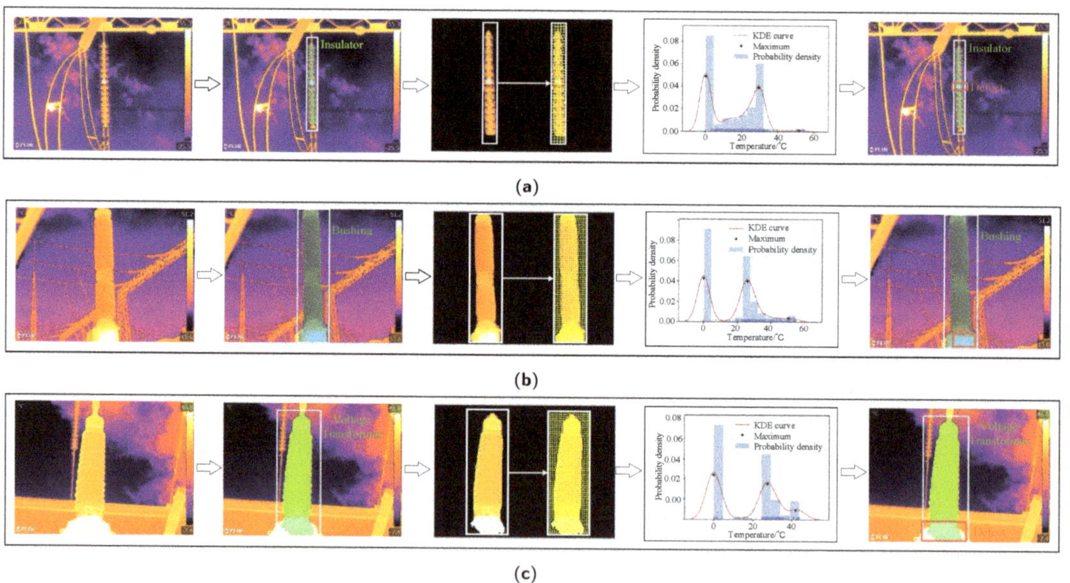

(a)

(b)

(c)

Figure 9. *Cont.*

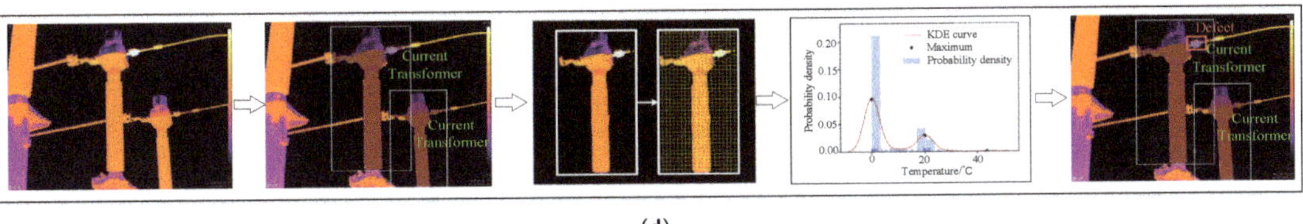

(d)

Figure 9. Process of the defect infrared image detection. (**a**) Insulator. (**b**) Bushing. (**c**) Voltage Transformer. (**d**) Current Transformer.

4. Discussion

In this paper, a two-level model is proposed for the problem of defect detection in substation equipment infrared images. On the basis of extracting substation equipment in the complex background through instance segmentation and superpixel segmentation methods, and realizing defect detection of substation equipment through temperature probability density distribution calculation and adaptive defect detection strategy. Compared with the traditional manual inspection, the proposed method can reduce the resources of labor and material; compared with the end-to-end deep learning method, the presented method in this paper does not require many defect datasets. The operating status of the substation equipment is closely relevant to the stability of the power system, which makes the defects detection of the substation equipment significant.

In the future, our research will not be limited to the substation equipment in this paper and will be applied to other electrical equipment. In fact, according to the characteristic of infrared thermal imaging, the majority of electrical equipment infrared images will show a certain temperature probability density distribution, which is the physical characteristic. The proposed method is based on this characteristic to detect defects precisely. Thus, based on this physical characteristic, we believe the method will be applicable to other cases where may occur defects in electric power, such as medical equipment, airplanes, and industrial equipment.

5. Conclusions

This study proposes a novel defect detection model named TDDM for infrared images of substation equipment. Considering the defective substation equipment infrared images are difficult to acquire, and the data-driven end-to-end model cannot be trained. Thus, the two-level defect detection method is presented. In the proposed TDDM, we take advantage of the fact that the instance segmentation has superior performance to extract the target in the redundant background. Meanwhile, the part of defect detection of TDDM is unsupervised and is not limited by the dataset. Furthermore, we evaluated the proposed model on the defect detection dataset, which accurately detects defects of substation equipment in infrared images. In the future, we would like to combine the RGB information to improve substation inspection tasks. In addition, the technology will be applied to more substation equipment.

Author Contributions: Conceptualization, B.L. and T.W.; methodology, T.W. and Z.H.; software, T.W.; validation, B.L., C.Y. and Y.Z.; formal analysis, Z.H.; investigation, T.W.; resources, Y.Z.; data curation, Y.Z.; writing—original draft preparation, T.W.; writing—review and editing, T.W., Z.H. and C.Y.; visualization, T.W.; supervision, Y.Z.; project administration, B.L.; funding acquisition, Y.Z. All authors have read and agreed to the published version of the manuscript.

Funding: This work was supported in part by the National Natural Science Foundation of China (NSFC) under grant number U21A20486, 61871182, by the Natural Science Foundation of Hebei Province of China under grant number F2020502009, F2021502008, F2021502013.

Institutional Review Board Statement: Not applicable.

Informed Consent Statement: Not applicable.

Data Availability Statement: Not applicable.

Acknowledgments: Special thanks are given to the Department of Automation and North China Electric Power University.

Conflicts of Interest: The authors declare no conflict of interest. The funders had no role in the design of the study; in the collection, analyses or interpretation of data; in the writing of the manuscript; or in the decision to publish the results.

Abbreviations

The following abbreviations are used in this manuscript:

DCNN	Deep Convolutional Neural Network
TDDM	Two-level Defect Detection Model
UHF	Ultra-high Frequency
FDR	frequency Domain Reflectometry
SNR	Signal-noise-ratio
TLD	Transmission Line Detection
NMS	Non-maximum Suppression
SLIC	Simple Linear Iterative Clustering
KDE	Kernel Density Estimation
mAP	Mean Aaverage precision
mAR	Mean Aaverage Recall
mRN	Mean Running Time
SSA	Superpixel Segmentation Algorithm
TP	True Positive
FP	False Positive
FN	False Negative

References

1. Han, S.; Yang, F.; Jiang, H.; Yang, G.; Zhang, N.; Wang, D. A smart thermography camera and application in the diagnosis of electrical equipment. *IEEE Trans. Instrum. Meas.* **2021**, *70*, 1–8. [CrossRef]
2. Wang, H.; Zhou, B.; Zhang, X. Research on the remote maintenance system architecture for the rapid development of smart substation in China. *IEEE Trans. Power Deliv.* **2017**, *33*, 1845–1852. [CrossRef]
3. Maina, R.; Tumiatti, V.; Pompili, M.; Bartnikas, R. Dielectric loss characteristics of copper-contaminated transformer oils. *IEEE Trans. Power Deliv.* **2010**, *25*, 1673–1677. [CrossRef]
4. Ozawa, J.; Shindo, K.; Saruta, H.; Yamashita, M.; Takahashi, E. Ultra high frequency electromagnetic wave detector for diagnostic of metal clad switchgear insulation. *IEEE Trans. Power Deliv.* **1994**, *9*, 675–679. [CrossRef]
5. Kwon, G.Y.; Lee, C.K.; Lee, G.S.; Lee, Y.H.; Chang, S.J.; Jung, C.K.; Kang, J.W.; Shin, Y.J. Offline fault localization technique on HVDC submarine cable via time–frequency domain reflectometry. *IEEE Trans. Power Deliv.* **2017**, *32*, 1626–1635. [CrossRef]
6. Zheng, H.; Sun, Y.; Liu, X.; Djike, C.L.T.; Li, J.; Liu, Y.; Ma, J.; Xu, K.; Zhang, C. Infrared image detection of substation insulators using an improved fusion single shot multibox detector. *IEEE Trans. Power Deliv.* **2020**, *36*, 3351–3359. [CrossRef]
7. Wang, B.; Dong, M.; Ren, M.; Wu, Z.; Guo, C.; Zhuang, T.; Pischler, O.; Xie, J. Automatic fault diagnosis of infrared insulator images based on image instance segmentation and temperature analysis. *IEEE Trans. Instrum. Meas.* **2020**, *69*, 5345–5355. [CrossRef]
8. Wang, Y. Improved OTSU and adaptive genetic algorithm for infrared image segmentation. In Proceedings of the 2018 Chinese Control And Decision Conference (CCDC), Shenyang, China, 9–11 June 2018; pp. 5644–5648.
9. Dongmei, W.; Ruyi, W.; Lihua, L. Automatic Detection of Oil Level of Transformer Oil Conservator Based on Infrared Image Segmentation Technology. In Proceedings of the 2011 Fourth International Conference on Intelligent Computation Technology and Automation, Shenzhen, China, 28–29 March 2011; Volume 1, pp. 596–599.
10. Niu, H.; Guo, S.; Xu, T.; Song, T.; Xu, L. Infrared image edge extraction of cable terminal based on improved eight direction Sobel operator. In Proceedings of the 2018 International Conference on Power System Technology (POWERCON), Guangzhou, China, 6–8 November 2018; pp. 3295–3300.
11. Canny, J. A computational approach to edge detection. *IEEE Trans. Pattern Anal. Mach. Intell.* **1986**, *PAMI-8*, 679–698. [CrossRef]
12. Liu, S.; Qi, X.; Shi, J.; Zhang, H.; Jia, J. Multi-scale patch aggregation (mpa) for simultaneous detection and segmentation. In Proceedings of the IEEE Conference on Computer Vision and Pattern Recognition, Las Vegas, NV, USA, 27–30 June 2016; pp. 3141–3149.
13. Ma, J.; Qian, K.; Zhang, X.; Ma, X. Weakly supervised instance segmentation of electrical equipment based on RGB-T automatic annotation. *IEEE Trans. Instrum. Meas.* **2020**, *69*, 9720–9731. [CrossRef]

4. Shu, J.; He, J.; Li, L. MSIS: Multispectral instance segmentation method for power equipment. *Comput. Intell. Neurosci.* **2022**, *2022*, 2864717. [CrossRef]
5. Cordts, M.; Omran, M.; Ramos, S.; Rehfeld, T.; Enzweiler, M.; Benenson, R.; Franke, U.; Roth, S.; Schiele, B. The cityscapes dataset for semantic urban scene understanding. In Proceedings of the IEEE Conference on Computer Vision and Pattern Recognition, Las Vegas, NV, USA, 27–30 June 2016; pp. 3213–3223.
6. Gupta, S.; Girshick, R.; Arbeláez, P.; Malik, J. Learning rich features from RGB-D images for object detection and segmentation. In Proceedings of the European Conference on Computer Vision, Zurich, Switzerland, 6–12 September 2014; pp. 345–360.
7. Hou, H.Q.; Liu, Y.J.; Lan, J.; Liu, L. Adaptive Fuzzy Fixed time Time-varying Formation Control for Heterogeneous Multi-agent Systems with Full State Constraints. *IEEE Trans. Fuzzy Syst.* **2022**, 1–10. [CrossRef]
8. Li, J.; Luo, G.; Cheng, N.; Yuan, Q.; Wu, Z.; Gao, S.; Liu, Z. An end-to-end load balancer based on deep learning for vehicular network traffic control. *IEEE Internet Things J.* **2018**, *6*, 953–966. [CrossRef]
9. Siheng, X.; Yadong, L.; Rui, X.; Ying, D.; Zihan, C.; Yingjie, Y.; Xiuchen, J. Power equipment recognition method based on mask R-CNN and bayesian context network. In Proceedings of the 2020 IEEE Power & Energy Society General Meeting (PESGM), Montreal, QC, Canada, 2–6 August 2020; pp. 1–5.
20. Ling, Z.; Zhang, D.; Qiu, R.C.; Jin, Z.; Zhang, Y.; He, X.; Liu, H. An accurate and real-time method of self-blast glass insulator location based on faster R-CNN and U-net with aerial images. *CSEE J. Power Energy Syst.* **2019**, *5*, 474–482.
21. Li, B.; Chen, C.; Dong, S.; Qiao, J. Transmission line detection in aerial images: An instance segmentation approach based on multitask neural networks. *Signal Process. Image Commun.* **2021**, *96*, 116278. [CrossRef]
22. Dey, D.; Chatterjee, B.; Dalai, S.; Munshi, S.; Chakravorti, S. A deep learning framework using convolution neural network for classification of impulse fault patterns in transformers with increased accuracy. *IEEE Trans. Dielectr. Electr. Insul.* **2017**, *24*, 3894–3897. [CrossRef]
23. Hui, Z.; Fuzhen, H. An intelligent fault diagnosis method for electrical equipment using infrared images. In Proceedings of the 2015 34th Chinese Control Conference (CCC), Hangzhou, China, 28–30 July 2015; pp. 6372–6376.
24. Liu, Y.; Pei, S.; Fu, W.; Zhang, K.; Ji, X.; Yin, Z. The discrimination method as applied to a deteriorated porcelain insulator used in transmission lines on the basis of a convolution neural network. *IEEE Trans. Dielectr. Electr. Insul.* **2017**, *24*, 3559–3566. [CrossRef]
25. Wen, L.; Li, X.; Gao, L.; Zhang, Y. A new convolutional neural network-based data-driven fault diagnosis method. *IEEE Trans. Ind. Electron.* **2017**, *65*, 5990–5998. [CrossRef]
26. Li, X.; Su, H.; Liu, G. Insulator defect recognition based on global detection and local segmentation. *IEEE Access* **2020**, *8*, 59934–59946. [CrossRef]
27. Wang, S.; Liu, Y.; Qing, Y.; Wang, C.; Lan, T.; Yao, R. Detection of insulator defects with improved resnest and region proposal network. *IEEE Access* **2020**, *8*, 184841–184850. [CrossRef]
28. Zhang, H.; Luo, G.; Tian, Y.; Wang, K.; He, H.; Wang, F.Y. A virtual-real interaction approach to object instance segmentation in traffic scenes. *IEEE Trans. Intell. Transp. Syst.* **2020**, *22*, 863–875. [CrossRef]
29. He, K.; Gkioxari, G.; Dollár, P.; Girshick, R. Mask r-cnn. In Proceedings of the IEEE International Conference on Computer Vision, Venice, Italy, 22–29 October 2017; pp. 2961–2969.
30. Huang, Z.; Huang, L.; Gong, Y.; Huang, C.; Wang, X. Mask scoring r-cnn. In Proceedings of the IEEE/CVF Conference on Computer Vision and Pattern Recognition, Long Beach, CA, USA, 15–20 June 2019; pp. 6409–6418.
31. Bolya, D.; Zhou, C.; Xiao, F.; Lee, Y.J. Yolact: Real-time instance segmentation. In Proceedings of the IEEE/CVF International Conference on Computer Vision, Seoul, Korea, 27 October–2 November 2019; pp. 9157–9166.
32. Gao, N.; Shan, Y.; Wang, Y.; Zhao, X.; Yu, Y.; Yang, M.; Huang, K. Ssap: Single-shot instance segmentation with affinity pyramid. In Proceedings of the IEEE/CVF International Conference on Computer Vision, Seoul, Korea, 27 October–2 November 2019; pp. 642–651.
33. Liu, S.; Jia, J.; Fidler, S.; Urtasun, R. Sgn: Sequential grouping networks for instance segmentation. In Proceedings of the IEEE International Conference on Computer Vision, Venice, Italy, 22–29 October 2017; pp. 3496–3504.
34. Wang, X.; Kong, T.; Shen, C.; Jiang, Y.; Li, L. Solo: Segmenting objects by locations. In Proceedings of the European Conference on Computer Vision, Glasgow, UK, 23–28 August 2020; pp. 649–665.
35. Salscheider, N.O. Featurenms: Non-maximum suppression by learning feature embeddings. In Proceedings of the 2020 25th International Conference on Pattern Recognition (ICPR), Milan, Italy, 10–15 January 2021; pp. 7848–7854.
36. Gaur, U.; Manjunath, B. Superpixel embedding network. *IEEE Trans. Image Process.* **2019**, *29*, 3199–3212. [CrossRef] [PubMed]
37. Liu, M.Y.; Tuzel, O.; Ramalingam, S.; Chellappa, R. Entropy rate superpixel segmentation. In Proceedings of the CVPR 2011, Colorado Springs, CO, USA, 20–25 June 2011; pp. 2097–2104.
38. Ren, X.; Malik, J. Learning a classification model for segmentation. In Proceedings of the Computer Vision, IEEE International Conference on IEEE Computer Society, Nice, France, 13–16 October 2003; Volume 2.
39. Li, Z.; Chen, J. Superpixel segmentation using linear spectral clustering. In Proceedings of the IEEE Conference on Computer Vision and Pattern Recognition, Boston, MA, USA, 7–12 June 2015; pp. 1356–1363.
40. Achanta, R.; Shaji, A.; Smith, K.; Lucchi, A.; Fua, P.; Süsstrunk, S. *Slic Superpixels*; Technical Report; EPFL: Lausanne, Switzerland, 2010.
41. Veksler, O.; Boykov, Y.; Mehrani, P. Superpixels and supervoxels in an energy optimization framework. In Proceedings of the European Conference on Computer Vision, Crete, Greece, 5–11 September 2010; pp. 211–224.

42. Levinshtein, A.; Stere, A.; Kutulakos, K.N.; Fleet, D.J.; Dickinson, S.J.; Siddiqi, K. Turbopixels: Fast superpixels using geometric flows. *IEEE Trans. Pattern Anal. Mach. Intell.* **2009**, *31*, 2290–2297. [CrossRef] [PubMed]
43. Liu, Z.; Shi, R.; Shen, L.; Xue, Y.; Ngan, K.N.; Zhang, Z. Unsupervised salient object segmentation based on kernel density estimation and two-phase graph cut. *IEEE Trans. Multimed.* **2012**, *14*, 1275–1289. [CrossRef]
44. Felzenszwalb, P.F.; Huttenlocher, D.P. Efficient graph-based image segmentation. *Int. J. Comput. Vis.* **2004**, *59*, 167–181. [CrossRef]
45. Fulkerson, B.; Soatto, S. Really quick shift: Image segmentation on a GPU. In Proceedings of the European Conference on Computer Vision, Crete, Greece, 5–11 September 2010; pp. 350–358.
46. Neubert, P.; Protzel, P. Compact watershed and preemptive slic: On improving trade-offs of superpixel segmentation algorithms. In Proceedings of the 2014 22nd International Conference on Pattern Recognition, Stockholm, Sweden, 24–28 August 2014; pp. 996–1001.
47. Ban, Z.; Liu, J.; Cao, L. Superpixel segmentation using Gaussian mixture model. *IEEE Trans. Image Process.* **2018**, *27*, 4105–4117. [CrossRef]

Article

Zero-Shot Image Classification Method Based on Attention Mechanism and Semantic Information Fusion

Yaru Wang [1], Lilong Feng [1], Xiaoke Song [1], Dawei Xu [1,2,*] and Yongjie Zhai [1]

1 Department of Automation, North China Electric Power University, Baoding 071003, China
2 State Key Laboratory of Management and Control for Complex Systems, Institute of Automation, Chinese Academy of Sciences, Beijing 100190, China
* Correspondence: xudawei@ncepu.edu.cn; Tel.: +86-176-2781-0027

Abstract: The zero-shot image classification (ZSIC) is designed to solve the classification problem when the sample is very small, or the category is missing. A common method is to use attribute or word vectors as a priori category features (auxiliary information) and complete the domain transfer from training of seen classes to recognition of unseen classes by building a mapping between image features and a priori category features. However, feature extraction of the whole image lacks discrimination, and the amount of information of single attribute features or word vector features of categories is insufficient, which makes the matching degree between image features and prior class features not high and affects the accuracy of the ZSIC model. To this end, a spatial attention mechanism is designed, and an image feature extraction module based on this attention mechanism is constructed to screen critical features with discrimination. A semantic information fusion method based on matrix decomposition is proposed, which first decomposes the attribute features and then fuses them with the extracted word vector features of a dataset to achieve information expansion. Through the above two improvement measures, the classification accuracy of the ZSIC model for unseen images is improved. The experimental results on public datasets verify the effect and superiority of the proposed methods.

Keywords: image classification; attention mechanism; matrix decomposition; attributes; word vectors

Citation: Wang, Y.; Feng, L.; Song, X.; Xu, D.; Zhai, Y. Zero-Shot Image Classification Method Based on Attention Mechanism and Semantic Information Fusion. *Sensors* 2023, 23, 2311. https://doi.org/10.3390/s23042311

Academic Editor: Paweł Pławiak

Received: 4 January 2023
Revised: 15 February 2023
Accepted: 16 February 2023
Published: 19 February 2023

1. Introduction

In recent years, deep learning algorithms have made rapid progress in the image recognition field, but they require significant human and material resources to obtain a sufficient quantity of manually annotated data [1]. In many practical applications, a large quantity of labeled data is difficult to obtain, and the variety of objects is increasing, which requires the computer training process to constantly add new samples and new object types [2,3]. The problem of how to use computers and existing knowledge to classify and identify samples with insufficient or even completely missing label data has become a pressing problem. For this reason, ZSIC [4] was created. It is a technique that trains a learning model to predict and recognize data without class labels (unseen classes) based on some sample data with class labels (seen classes), supplemented by relevant common-sense information or a priori knowledge (auxiliary information) [5,6].

To achieve ZSIC, a popular strategy is to learn the mapping or embedding between the semantic space of classes and the visual space of images based on seen classes and the semantic description of each category. Semantic descriptions of categories usually include attributes [7], word vectors [8], gaze [9], and sentences [10]. At present, the embedded-based methods [11–15] are used to learn visual-to-semantic, semantic-to-visual, or latent intermedium space, so that visual and semantic embedding can be compared in shared space. Then, the unseen classes are classified by nearest neighbor search.

Most of the existing embedding methods, either based on end-to-end convolution neural networks or deep features, emphasize learning the embedding between global visual features and semantic vectors, which leads to two problems [16]. First, there are only slight differences between some features of seen and unseen classes. For some datasets, the inter-class difference is even smaller than the intra-class. Therefore, global image features cannot effectively represent fine-grained information, which is difficult to distinguish in semantic space. Second, compared to visual information, semantic information is not rich enough. The attribute features of categories are usually based on manual annotation, rely on professional knowledge, and are limited by the dimension of visual cognition. The dimension of attribute features is usually not high, and as intermediate auxiliary information, the amount of information is insufficient [17]. The word vectors are mostly obtained through models such as word2vec [18], GloVe [19], or fastText [20]. Relatively speaking, the word vectors may contain more noise and are difficult to combine with human prior knowledge; thus, their interpretability and discriminability are poor. Therefore, the imbalanced supervision from the semantic and visual space can make the learned mapping easily overfitting to seen classes. Inspired by the attention mechanism in the field of natural language processing, a few methods [16,21–23] introduce attention thinking into ZSIC. These methods learn regional embedding of different attributes or similarity measures based on attribute prototypes and learn to distinguish partial features, but they ignore the global features and the information imbalance of semantic and visual space.

Based on the above observation, this paper proposes an improved ZSIC model. The main contributions are as follows:

(1) A feature attention mechanism is designed, and an image feature extraction module based on the attention mechanism is built. The features in different regions of the image are assigned attention weights to distinguish the key and non-key local features, and then the local features are fused with the global features.
(2) A semantic information fusion module based on matrix decomposition is built. The matrix decomposition method is used to transform the binary features of attributes into continuous features and transform their dimensions to be the same as word vectors. In addition, attribute features are fused with word vector features to obtain more accurate and richer fused semantic features as a priori category features.
(3) The improved ZSIC model promotes the alignment of semantic information and visual features. Experiments on the public dataset show that the improved ZSIC model improves image classification accuracy.

2. Related Work

2.1. ZSIC Methods

Recent ZSIC methods focus on learning better visual–semantic embeddings. The core idea is to learn a mapping between the visual and attribute/semantic domains and transfer semantic knowledge from seen to unseen classes according to the similarity measure. Some methods [11,12,24,25] follow the visual-to-semantic mapping direction and align visual features and semantic information in semantic space. However, when high-dimensional visual features are mapped to a low-dimensional semantic space, the shrink of feature space would aggravate the hubness problem [26,27] that in some instances in the high-dimensional space becomes the nearest neighbors of a large number of instances. To tackle these problems, some methods [13,14,28–30] map semantic embedding to visual space and treat the projected results as class prototypes. Shigeto et al. [31] experimentally proved that the semantic-to-visual embedding is able to generate more compact and separative visual feature distribution with the one-to-many correspondence manner, thereby mitigating the hubness issue. Ji et al. [32] also follow the inverse mapping direction from semantic space to visual space and proposed a semantic-guided class imbalance learning model which alleviates the class-imbalance issue in ZSIC. In addition, for the class-imbalance issue, the generative models have been introduced to learn semantic-to-visual mapping to generate visual features of unseen classes [33–37] for data augmentation. Currently, the generative ZSIC is usually based on variational autoencoders (VAEs) [37], generative adversarial nets (GANs) [33], and generative flows [34]. However, the performance of this type of method greatly depends on the quality of generated visual features or images, which is difficult to guarantee, and the mode is prone to mode collapse. Furthermore, to alleviate the hubness issue, common space learning is also employed to learn a common representation space for interaction between visual and semantic domains [15,38,39]. However, these embedded-based models only use the global feature representation, ignoring the fine-grained details in the image, and the training results are not satisfied for the poorly identified features.

2.2. Attention Mechanism

The concept of attention was first introduced into natural language processing tasks. In particular, because soft attention is differentiable and can learn parameters by back-propagation of the model, it has been widely used and developed in computer vision tasks. Zhu et al. [40] applied an attention mechanism in the facial expression recognition task and proposed a cascade attention-based recognition network by a hybrid of the spatial attention mechanism and pyramid feature to improve the accuracy of facial expression recognition under uneven illumination or partial occlusion. Sun et al. and Liu et al. applied an attention mechanism in the semantic segmentation task of remote sensing images. They proposed a multi-attention-based UNet [41] and an attention-based residual encoder [42], respectively. Through channel attention and spatial attention, the capability of fine-grained features was improved. The above attention mechanism includes (i) feature aggregation and (ii) a combination of channel attention (global attention) and spatial attention (local attention), which are common branches of the attention mechanism. In addition, Obeso et al. [43] proved that the global and local attention mechanism in deep neural networks works well with the human visual attention mechanism. Inspired by the above works, several researchers incorporated an attention mechanism into models for ZSIC. For example, Yang et al. [16] proposed a semantic-aligned reinforced attention model to discover invariable features related to class-level semantic attributes from variable intra-class vision information, and thereby to avoid misalignment between visual information and semantic representations. Xu et al. [21] jointly learned discriminative global and local features using only class-level attributes to improve the attribute localization ability of image representation. Chen et al. [22] proposed an attribute-guided transformer network to enhance discriminative attribute localization by reducing the relative geometry relationships among the grid features. Yang et al. [23] proposed to learn prototypes via placeholders and proposed semantic-oriented fine-tuning for preliminary visual–semantic alignment. These methods locate salient regions according to semantic attributes and ignore meaningless information to promote the alignment between a visual space and a semantic space. Compared with these methods, we also consider the combination of local features and global features, as well as the imbalance of information in semantic and visual space.

3. Materials and Methods

The basic embedding-based ZSIC model framework is shown in Figure 1.

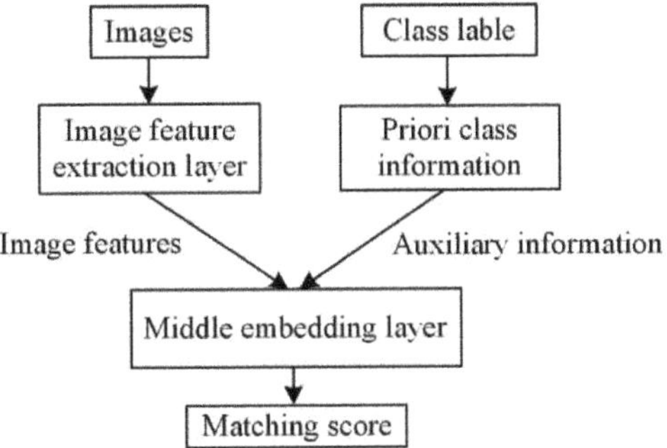

Figure 1. Basic embedding-based ZSIC model framework.

The image feature extraction layer uses a deep CNN to extract image features and input them to a middle embedding layer. A priori class information (auxiliary information) is usually attribute features or word vector features. In the middle embedding layer, the correlation between image features and a priori class information is calculated. Let the total number of seen classes be n and a priori class feature vector of the i-th seen class be β_i, whose dimension is m. In the training stage of the model, the images x_i belonging to the i-th seen class are input into the image feature extraction layer to extract m-dimensional image feature vectors α_{x_i}; α_{x_i} and β_i are input into the middle embedding layer, and a relationship similarity (α_{x_i}, β_i) between α_{x_i} and β_i is established to obtain the matching score. Cosine distance is used to calculate the matching score. Compared with the European distance, cosine distance is more consistent with the distance calculation form of the high-dimensional vector, and its formula is

$$\text{score} = \text{similarity}(\alpha_{x_i}, \beta_i) = \frac{\sum_{k=1}^{m} a_k b_k}{\sqrt{\sum_{k=1}^{m} a_k^2} \sqrt{\sum_{k=1}^{m} b_k^2}} \tag{1}$$

where $\alpha_{x_i} = [a_1, a_2, \ldots, a_m]$ and $\beta_i = [b_1, b_2, \ldots, b_m]$.

In order to match the image feature vectors and the prior class feature vectors belonging to the same class as closely as possible, that is, to maximize the matching score, the loss function is used as follows:

$$\text{loss} = -\frac{1}{n} \sum_{i=1}^{n} \frac{\alpha_{x_i} \cdot \beta_i}{\| \alpha_{x_i} \| \cdot \| \beta_i \|} \tag{2}$$

In the testing stage of the model, the image feature vectors of unseen classes are extracted through the feature extraction layer and then matched with the prior class feature vectors corresponding to each class in the middle embedding layer. When the matching score is the highest, the corresponding class is the prediction class of the input image.

Using the above model framework, the improved embedding-based ZSIC model is shown in Figure 2. Details are as follows.

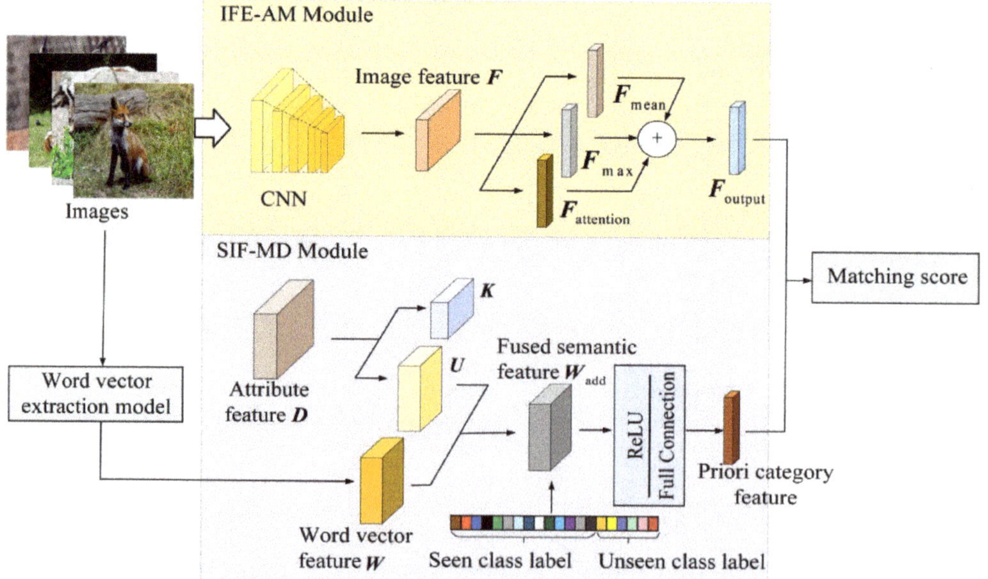

Figure 2. Improved ZSIC model.

3.1. IFE-AM Module

In ZSIC tasks, image features need to be matched with a priori class features, while image features extracted by CNN correspond to a whole image, so they lack discrimination. Therefore, an image feature extraction module based on an attention mechanism (IFE-AM) is constructed (as shown in Figure 2) to focus high-level image features on the key regions of the input image, in order to reduce the deviation from the priori class features and improve the degree of matching. The typical convolutional neural networks VGG-19 and ResNet-34 are taken as examples to illustrate the attention mechanism designed in this paper.

The flowchart of the spatial attention mechanism that weights the feature vector of each position is shown in Figure 3.

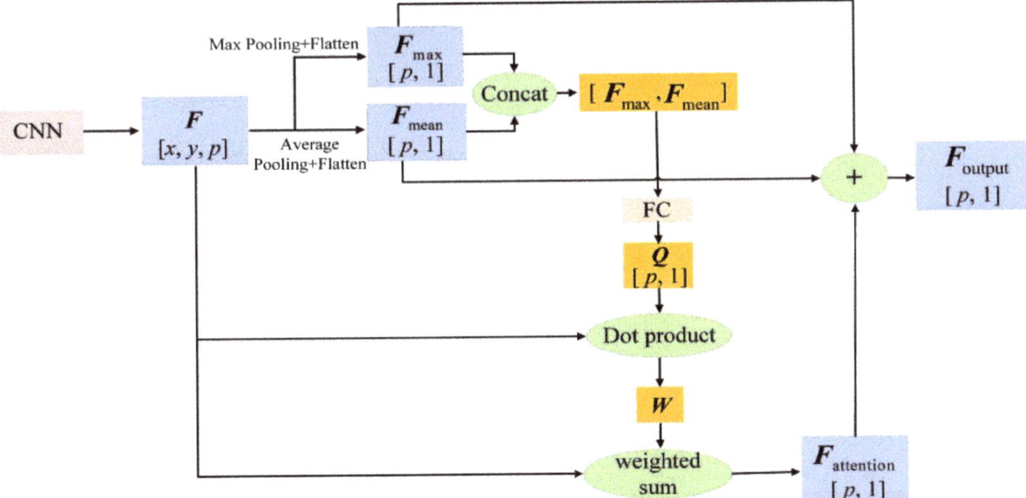

Figure 3. Flowchart of the attention mechanism.

Let the output features of the last layer of the CNN be F, with dimension $[x, y, p]$, which contains p channels. For F, set window $[x, y]$, and use max pooling and average pooling to obtain two p-dimensional feature vectors F_{max} and F_{mean}, respectively, and then concatenate them to obtain $[F_{max}, F_{mean}]$. Then, $[F_{max}, F_{mean}]$ is connected to the fully connected (FC) layer, the hidden layer unit is set as p, and a p-dimensional query vector Q is output for feature selection of the attention mechanism. The feature map of the i-th channel in F is recorded as f_i, $i = 1, 2, \ldots, p$, and its size is $x \times y$; the feature vector of the j-th position in F is recorded as l_j, $j = 1, 2, \ldots, x \times y$, and its size is $p \times 1$. Calculate the dot product of Q and l_j to obtain the feature weight w_j of the j-th position, and then use the softmax function for normalization to obtain the feature weight matrix W. The formula is as follows:

$$W = \text{softmax}\left(w_j\right) = \text{softmax}(\text{dot}(Q^T, l_j)) \tag{3}$$

The feature values at different positions in f_i are weighted and summed according to the weight matrix W, and $F_{attention}$ is output.

Finally, based on the idea of residual connection, the feature vectors F_{max}, F_{mean}, and $F_{attention}$ are summed to obtain the final output eigenvector F_{output}.

3.2. SIF-MD Module

ZSIC methods rely on prior class information to complete the transfer from seen classes to unseen classes, so accurate and informative class description information is the key. Currently, the commonly used a priori class description information includes attribute features and word vector features. In order to make the two types of a priori class description information complementary and improve the amount of information, a semantic information fusion module based on matrix decomposition (SIF-MD) is constructed, as shown in Figure 2.

Usually, the dimensions of manually set attribute information is small, and the attribute features are all binary features of 0 or 1, which are relatively sparse and independent; the dimensions of word vectors are relatively large, which are characterized by continuity between [−1, 1]. To carry out information fusion, the matrix decomposition method is used to transform the binary features of attributes into continuous features and transform their dimensions to be the same as word vectors. The architecture diagram of the matrix decomposition of attributes is shown in Figure 4.

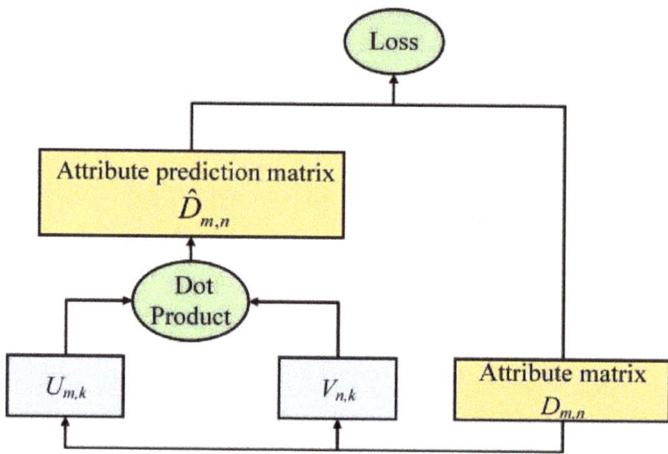

Figure 4. Architecture diagram of the matrix decomposition of attributes.

First, use attribute matrix D ($M \times N$) to represent n-dimensional attribute vectors of m classes, which is decomposed into U ($M \times K$) and V ($N \times K$) with the equation

$$D = UV^{\mathrm{T}} \tag{4}$$

where k is the dimension of the matrix decomposition. Make UV^{T} as close as possible to D, that is, fitting attribute feature D through matrix U and matrix V. The loss function is the mean squared error MSE (mean squared error) method:

$$\mathrm{loss} = \sum_{i=1}^{M} \sum_{j=1}^{N} \left(D_{i,j} - \hat{D}_{i,j}\right)^2 \tag{5}$$

$$\hat{D}_{i,j} = U_i V_j^{\mathrm{T}} \tag{6}$$

where U_i denotes the vector in the i-th row of matrix U, $i = 1, 2, \ldots, M$, and V_j denotes the vector in the j-th row of matrix V, $j = 1, 2, \ldots, N$.

To prevent overfitting, the L2 canonical term is added to Formula (5):

$$\mathrm{loss} = \sum_{i=1}^{m} \sum_{j=1}^{n} \left(D_{i,j} - \hat{D}_{i,j}\right)^2 + \lambda\left(\|U_i\|_1 + \|V_j\|_1\right) \tag{7}$$

Each row in U is a k-dimension vector, which matches the dimension of the word vector of the corresponding class. The matrix U and the word vector matrix $W(m \times k)$ are summed in certain weight proportions as fused semantic features W_{add}, which are given by

$$W_{add} = \alpha W + (1 - \alpha)U \tag{8}$$

where α is a parameter with a range of [0, 1]; W_{add} is a fused semantic feature, retaining the content of attribute features and word vector features.

4. Experiment Results

The experiment is based on the $4\times$ 1080Ti GPU server of Ubuntu16.04, the Python 3.6 virtual environment is built through Anaconda, and deep learning frameworks of TensorFlow1.2.0 and Keras2.0.6 are installed.

The top-1 accuracy and top-3 accuracy were used to evaluate the classification results of the zero-shot classification model on the test set. The training set and test set were randomly selected four times to obtain four groups of experimental results, and the average classification accuracy was recorded.

4.1. Dataset

The experiment was conducted based on the Animals with Attributes 2 (AwA2) [27] dataset. AwA2 is a public dataset for attribute-based classification and zero-shot learning, and it is publicly available at http://cvml.ist.ac.at/AwA2, accessed on 9 June 2017. The dataset contains 37,322 images and 50 animal classes, and each class has an 85-dimensional attribute vector. It is a coarse-grained dataset that is medium-scale in terms of the number of images and small-scale in terms of the number of classes. In experiments, we followed the standard zero-shot split proposed in reference [9], that is, 40 classes for training and 10 classes for testing. The training set and test set do not intersect. Among the training set, 13 classes were randomly selected for validation to perform a hyperparameter search.

4.2. Ablation Experiment of IFE-AM Model

According to the model structure shown in Figure 2, the experiments were conducted with the representative VGG-19 and ResNet-34 as the backbone networks, which are called VGG-A and ResNet-A, respectively. The image features were extracted by the pre-improved and improved networks, and the attribute features of the dataset were used to conduct experiments.

4.2.1. Training Loss and Classification Accuracy

When the model is trained, the training loss is calculated according to Formula (2). Figure 5 shows the change curves of the training loss (train_loss) corresponding to different feature extraction networks.

Table 1 shows the epochs required for training and train_loss values corresponding to different feature extraction networks, as well as the classification accuracy (top-1 and top-3) of the test set.

Figure 5 and Table 1 show that the train_loss of the ResNet-34 model decreases faster than the VGG-19 model. The final train_loss of the VGG-19 and ResNet-34 models tends to be stable, but the train_loss of the ResNet-34 model is lower. From the decreasing trend in train_loss, the train_loss of the VGG-19 model fluctuates greatly, and the decreasing process of train_loss of the ResNet-34 model is more stable. The ResNet-A model is also superior to the VGG-A model in decreasing speed and the stability of train_loss. This shows that the ResNet-34 model with residual connections can realize matching between image features and prior class features faster, better, and more stably. In addition, for both the VGG-A model and ResNet-A model, although their train_loss overall declines slightly slower, their required training epoch and loss value after stabilization are significantly lower than those of the original VGG-19 and ResNet-34 networks. This shows that the IFE-AM module proposed in this paper, as a feature-weighted focusing strategy, improves the model's ability to capture image features in space, thus realizing further fitting of deep features; additionally, the attention mechanism is based on the method of weighted information fusion, which makes the acquisition and update of information more stable, thus achieving a faster and more stable fitting effect.

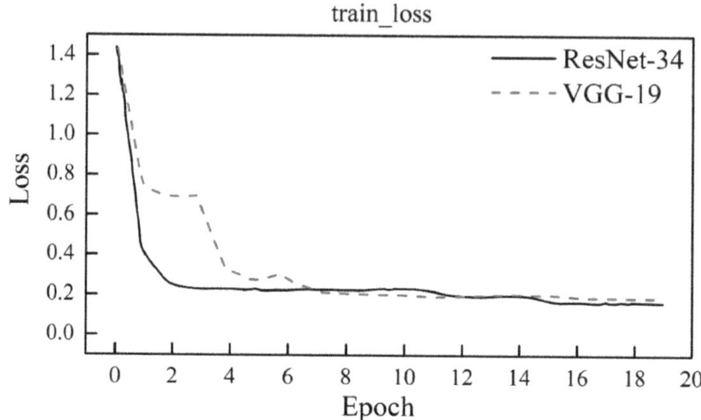

(**a**) Change curve of train_loss corresponding to VGG-19 and ResNet-34

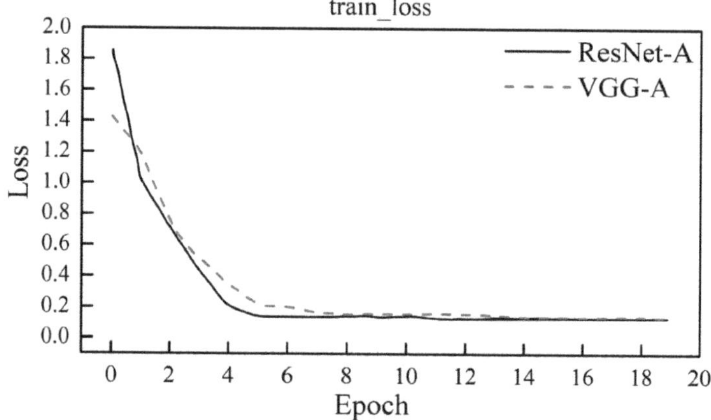

(**b**) Change curve of train_loss corresponding to VGG-A and ResNet-A

Figure 5. Change curves of train_loss.

Table 1. Test results.

Feature Extraction Network	IFE-AM	Epochs	Train_Loss	Top-1 (%)	Top-3 (%)
VGG-19		17	0.174	40.1	53.1
ResNet-34		16	0.155	41.7	56.1
VGG-A	√	13	0.147	43.2	60.9
ResNet-A	√	5	0.139	43.3	63.9

For the image classification results of the test set, the top-1 and top-3 of the ResNet-34 model are all larger than those of the VGG-19 model, which shows that its residual structure has a good effect on the fitting of deep image features. The top-1 and top-3 of the ResNet-A model are higher than those of the VGG-19 and ResNet-34 models without the attention mechanism, which shows that the attention mechanism can focus the features of spatial attention and effectively improve the generation of image features and the matching effect with prior class features. The accuracies of VGG-A and ResNet-A are similar, but the top-3 of ResNet-A is significantly improved, which shows that the ResNet-A model can obtain more accurate image features in high-dimensional space, making the distance between classes farther, the distance within classes closer, and the matching effect with semantic features better.

4.2.2. Feature Segmentation

According to the model shown in Figure 4, for VGG-A and ResNet-A, the image feature $F_{output} = F_{max} + F_{mean} + F_{attention}$ is split, and F_{max}, F_{mean} and $F_{attention}$ are, respectively, output to the next layer for comparison with F_{output}. The accuracy of the final image classification is shown in Tables 2 and 3.

Table 2. Comparison of different image features in the VGG-A model.

Image Features	Attention	Feature Fusion	Top-1 (%)	Top-3 (%)
F_{max}			39.9	45.0
F_{mean}			40.3	51.1
$F_{attention}$	√		40.9	51.9
F_{output}	√	√	42.3	60.9

Table 3. Comparison of different image features in the ResNet-A model.

Image Features	Attention	Feature Fusion	Top-1 (%)	Top-3 (%)
F_{max}			39.1	41.1
F_{mean}			41.7	56.1
$F_{attention}$	√		42.9	61.1
F_{output}	√	√	43.3	63.9

As shown in Tables 2 and 3, the image classification results of the improved ResNet-A model based on the attention mechanism are better than those of the VGG-A model. Whether it is the VGG-A or ResNet-A model, the image classification accuracy corresponding to different image features satisfies $F_{output} > F_{attention} > F_{mean} > F_{max}$, which verifies the effect of image feature extraction based on the spatial attention mechanism. Inspired by the idea of residual connection, the three features are superposed to obtain F_{output}, which fuses the information of different features and finally obtains the optimal image classification result.

4.3. Ablation Experiment of SIF-MD Module

Since the above experiments verified that ResNet-A and F_{output} are better, the following further experiments are conducted on these bases. Three models of word2vec, GloVe, and fastText were used to extract the word vector features of each class in the dataset, with a dimension of 256. The attribute features of the dataset were decomposed according to Formulas (4)–(7), and the loss threshold value was set as 0.1. Then, the decomposed attributes were weighted and fused with word vector features extracted by word2vec, GloVe, and fastText, respectively, according to Formula (8). The fusion parameter α was set as [0, 1] and the step size as 0.1.

The image classification experiment of the test set was repeated five times, and the average value of the top-1 was taken. The experimental results corresponding to different word vectors and different fusion parameters α are shown in Table 4. Figure 6 more intuitively shows the changing trend of top-1 accuracy with α when different word vectors are used as auxiliary information.

Table 4. Image classification top-1 accuracy of the test set.

Word Vector	α										
	0	0.1	0.2	0.3	0.4	0.5	0.6	0.7	0.8	0.9	1
word2vec	43.1	43.1	43.1	43.3	43.7	43.8	43.8	44.0	44.3	44.5	44.2
GloVe	43.1	44.3	44.6	44.6	44.6	44.7	45.0	45.1	45.8	45.3	44.7
fastText	43.1	43.0	43.3	43.6	43.2	42.8	42.5	42.5	42.2	42.2	42.1

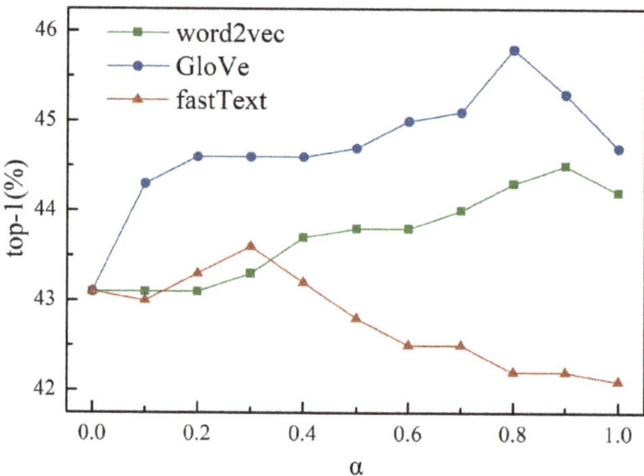

Figure 6. Changing trend of top-1 accuracy of image classification.

As shown in Figure 6, the top-1 accuracy of the word vector extracted by GloVe as prior class features is significantly higher than that extracted by word2vec or fastText. As shown in Table 4, when $\alpha = 0$, that is, only the attribute features are used as the prior class feature, the top-1 accuracy of image classification is 43.1%. When $\alpha = 1$, that is, only word vectors are used as prior class features, the top-1 accuracies corresponding to word2vec and GloVe are 44.2% and 44.7%, respectively, which are better than the results when only attribute features are used, while the top-1 accuracy corresponding to fastText is lower than the results when only attribute features are used. For the word vectors extracted by word2vec, GloVe, and fastText, the fusions with attribute feature all have positive effects. For the word2vec word vector, when the fusion weight $\alpha = 0.8$ and 0.9, the top-1 accuracy is 1.2% and 1.4% higher than that of the attribute vector only and 0.1% and 0.3% higher than that of the word vector only, respectively. For the fastText word vector, when the fusion weight $\alpha = 0.2$, 0.3, and 0.4, the top-1 accuracy is 0.2%, 0.5%, and 0.1% higher than that of the attribute vector only and 1.2%, 1.5%, and 1.1% higher than that of the word vector only, respectively. For the GloVe word vector, when the fusion weight $\alpha = 0.6$, 0.7, 0.8, and 0.9, the top-1 accuracy is 1.9%, 2.0%, 2.7%, and 2.2% higher than that of the attribute vector only and 0.3%, 0.4%, 1.1%, and 0.6% higher than that of the word vector only, respectively. The results show that it is meaningful to fuse attribute features and the word vector features.

5. Discussions

To verify the effectiveness of the method proposed, the method is compared with the baseline model and existing classical models. The baseline model only uses the deep learning network ResNet-34 or VGG-19 to extract image features and uses attributes or word vectors as auxiliary information. The results of the comparative experiment are shown in Table 5 and Figure 7. In the table, "ResNet-34 + attribute" refers to the model that uses ResNet-34 to extract image features and uses attributes as auxiliary information. The image classification results were evaluated with top-1 accuracy. The experimental results of IAP, CONSE, and CMT adopt the results given in references [27,31]. The dataset and the splits of the training set and test set in the experiments of all methods are the same as that of our method, and no methods were pre-trained by large datasets (such as ImageNet).

Table 5. Image classification results of different methods.

	Method	Top-1 (%)
1	ResNet-34 + attribute	41.7
2	ResNet-34 + word2vec	42.3
3	ResNet-34 + GloVe	42.7
4	ResNet-34 + fastText	40.6
5	VGG-19 + attribute	40.1
6	VGG-19 + word2vec	40.4
7	VGG-19 + GloVe	41.2
8	VGG-19 + fastText	39.9
9	IAP	35.9
10	CONSE	44.5
11	CMT	37.9
12	ours	45.8

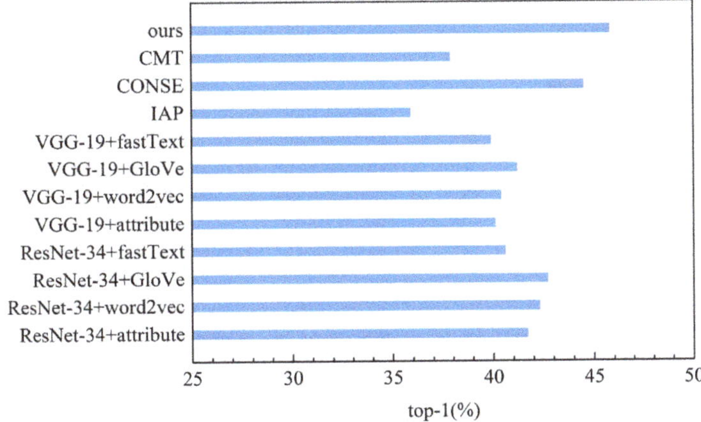

Figure 7. Top-1 accuracy comparison of different methods.

As shown in Table 5 and Figure 7, for the baseline model, the top-1 accuracy of the model using ResNet-34 to extract image features is higher than that of the model using the VGG-19 network; the top-1 accuracy of the model using word vectors extracted by word2vec or GloVe as auxiliary information is higher than that of the model using attributes; and the top-1 accuracy of the "ResNet-34 + GloVe" method is the highest, with a value of 42.7%. The top-1 accuracy of our method is 3.1% higher than that of the "ResNet-34 + GloVe" method. For existing classical methods, IAP detects unseen classes based on attribute transfer between classes, the attribute features are limited by the dimension of visual cognition, and the amount of information is insufficient. CONSE uses CNN to extract image features without distinguishing the importance of different regional features, and only uses word vectors extracted by word2vec as auxiliary information. CMT uses Sparse Coding to extract image features and uses a neural network architecture to learn the word vectors of categories. Although more semantic word representations are learned by using local and global contexts, the discrimination of word vectors is poor, and the imbalanced supervision between semantic features and visual features is still large. Our method assigns attention weights to different regions of the image through the SIF-MD module and strengthens the key features highly related to semantic information. In addition, it alleviates the imbalanced supervision issue between semantic features and visual features through IFE-AM module. These improvements promote the alignment of visual features and semantic information and make the matching degree of the two higher, which is very important for ZSIC. Thus, the top-1 accuracy of our method is 9.9% higher than IAP, 1.3% higher than CONSE, and 7.9% higher than CMT. The above experimental results prove the effectiveness of our method.

6. Conclusions

To improve the accuracy of the ZSIC model based on embedded space, the IFE-AM model and SIF-MD module are constructed in this paper. After the existing CNN is used to extract the image feature map, the max pooling, average pooling, and spatial attention methods are used to obtain three feature vectors, and then they are fused as the final image features. The attribute matrix of the dataset is decomposed to match its dimensions with the extracted word vector, and then the attribute and word vector are weighted and fused as auxiliary information of the improved ZSIC model.

Experiments were conducted on a public dataset. First, the ablation experiment of the IFE-AM model was carried out. The experimental results show that the top-1 and top-3 accuracies corresponding to ResNet-A are 1.6% and 7.8% higher than those of ResNet-34, respectively; the top-1 and top-3 accuracies corresponding to VGG-A are 3.1% and 7.8% higher than those of VGG-19, respectively. Then, the ablation experiment of the SIF-MD module was carried out. The experimental results show that the top-1 accuracies of using fused semantic information as auxiliary information are significantly higher than that of using attribute or word vector alone. Third, comparative experiments were carried out, and the results show that the accuracy of the proposed method is significantly higher than the baseline method and several existing classical methods.

For different types of semantic information, the fusion parameter is not fixed and needs to be determined by experiments. How to derive the value of the fusion parameter in theory is our future work. A small- to medium-sized dataset is considered in our work, and larger data scenarios will be explored in the future.

Author Contributions: Conceptualization, Y.W. and D.X.; methodology, Y.W. and L.F.; software, L.F.; validation, L.F. and X.S.; data curation, X.S.; writing—original draft preparation, L.F.; writing—review and editing, X.S. and Y.Z.; supervision, Y.Z. All authors have read and agreed to the published version of the manuscript.

Funding: This paper was supported by the following projects: Natural Science Foundation Youth Science Fund Project of Hebei Province (No. F2021502008): "Research on Zero-shot Fault Detection Method for Transmission Lines with Domain Knowledge"; the General Project of the Fundamental Research Funds for the Central Universities (No. 2021MS081): "Research on Transmission Lines Fault Detection Method with Multimodal Information"; and Open Research Fund of The State Key Laboratory for Management and Control of Complex Systems (No. 20220102): "Research on Interactive Control Method of Snake-Like Manipulator".

Institutional Review Board Statement: Not applicable.

Informed Consent Statement: Not applicable.

Data Availability Statement: Not applicable.

Acknowledgments: Special thanks are given to North China Electric Power University.

Conflicts of Interest: The authors declare no conflict of interest. The funders had no role in the design of the study, in the collection, analyses or interpretation of data, in the writing of the manuscript, or in the decision to publish the results.

Abbreviations

The following abbreviations are used in this manuscript:

ZSIC	Zero-shot image classification
CNNs	Convolutional neural networks
IFE-AM	Image feature extraction module based on an attention mechanism
SIF-MD	Semantic information fusion module based on matrix decomposition
AwA2	Animals with Attributes 2
FC	Fully connect

References

1. Lecun, Y.; Bengio, Y.; Hinton, G. Deep learning. *Nature* **2015**, *521*, 436. [CrossRef]
2. Sun, X.; Gu, J.; Sun, H. Research progress of zero-shot learning. *Appl. Intell.* **2021**, *51*, 3600–3614. [CrossRef]
3. Li, L.W.; Liu, L.; Du, X.H.; Wang, X.; Zhang, Z.; Zhang, J.; Liu, J. CGUN-2A: Deep Graph Convolutional Network via Contrastive Learning for Large-Scale Zero-Shot Image Classification. *Sensors* **2022**, *22*, 9980. [CrossRef]
4. Palatucci, M.; Pomerleau, D.; Hinton, G.E. Zero-shot learning with semantic output codes. In Proceedings of the Advances in Neural Information Processing Systems, Vancouver, BC, Canada, 7–10 December 2009; pp. 1410–1418.
5. Li, Z.; Chen, Q.; Liu, Q. Augmented semantic feature based generative network for generalized zero-shot learning. *Neural Netw.* **2021**, *143*, 1–11. [CrossRef]
6. Ohashi, H.; Al-Naser, M.; Ahmed, S.; Nakamura, K.; Sato, T.; Dengel, A. Attributes' Importance for Zero-Shot Pose-Classification Based on Wearable Sensors. *Sensors* **2018**, *18*, 2485. [CrossRef]
7. Wu, L.; Wang, Y.; Li, X.; Gao, J. Deep attention-based spatially recursive networks for fine-grained visual recognition. *IEEE Trans. Cybern.* **2018**, *49*, 1791–1802. [CrossRef]
8. Krizhevsky, A.; Sutskever, I.; Hinton, G.E. ImageNet classification with deep convolutional neural networks. In Proceedings of the Advances In Neural Information Processing Systems, Lake Tahoe, NV, USA, 3–6 December 2012; pp. 1097–1105.
9. Lampert, C.; Nickisch, H.; Harmeling, S. Attribute-based classification for zero-shot visual object categorization. *IEEE Trans. Pattern Anal. Mach. Intell.* **2014**, *36*, 453–465. [CrossRef]
10. He, K.; Zhang, X.; Ren, S.; Sun, J. Deep residual learning for image recognition. In Proceedings of the IEEE Conference on Computer Vision and Pattern Recognition, Las Vegas, NV, USA, 27–30 June 2016; pp. 770–778.
11. Xu, W.J.; Xian, Y.Q.; Wang, J.N.; Schiele, B.; Akata, Z. Attribute prototype network for zero-shot learning. *Neural Inf. Process. Syst.* **2020**, *33*, 21969–21980.
12. Xie, G.S.; Liu, L.; Jin, X.B.; Zhu, F.; Zhang, Z.; Qin, J.; Yao, Y.Z.; Shao, L. Attentive region embedding network for zero-shot learning. In Proceedings of the IEEE/CVF Conference on Computer Vision and Pattern Recognition, Long Beach, CA, USA, 16–17 June 2019; pp. 9376–9385.
13. Li, K.; Min, M.R.; Fu, Y. Rethinking zero-shot learning: A conditional visual classification perspective. In Proceedings of the IEEE/CVF International Conference on Computer Vision, Seoul, Republic of Korea, 27 October–2 November 2019; pp. 3583–3592.
14. Zhang, L.; Xiang, T.; Gong, S. Learning a deep embedding model for zero-shot learning. In Proceedings of the IEEE Conference on Computer Vision and Vattern Recognition, Honolulu, HI, USA, 21–26 July 2017; pp. 2021–2030.
15. Chen, S.M.; Xie, G.S.; Liu, Y.Y.; Peng, Q.M.; Sun, B.G.; Li, H.; You, X.G.; Ling, S. Hsva: Hierarchical semantic-visual adaptation for zero-shot learning. *Neural Inf. Process. Syst.* **2021**, *34*, 16622–16634.
16. Zhu, Y.Z.; Tang, Z.; Peng, X.; Elgammal, A. Semantic-guided multi-attention localization for zero-shot learning. In Proceedings of the Advances in Neural Information Processing Systems, Vancouver, BC, Canada, 8–14 December 2019; Volume 32.
17. Jayaraman, D.; Kristen, G. Zero-shot recognition with unreliable attributes. In Proceedings of the International Conference on Neural Information Processing Systems, Montreal, QC, USA, 8–13 December 2014; pp. 3464–3472.
18. Mikolov, T.; Sutskever, I.; Chen, K.; Corrado, G.S.; Dean, J. Distributed representations of words and phrases and their compositionality. In Proceedings of the Advances in Neural Information Processing Systems, Lake Tahoe, NV, USA, 5–8 December 2013; pp. 3111–3119.
19. Pennington, J.; Socher, R.; Manning, C.D. Glove: Global vectors for word representation. In Proceedings of the 2014 Conference on Empirical Methods in Natural Language Processing, Doha, Qatar, 25–29 October 2014; pp. 1532–1543.
20. Joulin, A.; Grave, E.; Bojanowski, P.; Mikolov, T. Bag of tricks for efficient text classification. *arXiv* **2016**, arXiv:1607.01759.
21. Xu, W.; Xian, Y.; Wang, J.; Schiele, B.; Akata, Z. Attribute prototype net-work for zeroshot learning. *arXiv* **2020**, arXiv:2008.08290.
22. Chen, S.; Hong, Z.; Liu, Y.; Xie, G.S.; Sun, B.; Li, H.; Peng, Q.; Lu, K.; You, X. Transzero: Attribute-guided transformer for zero-shot learning. *arXiv* **2021**, arXiv:2112.01683. [CrossRef]
23. Yang, Z.; Liu, Y.; Xu, W.; Huang, C.; Zhou, L.; Tong, C. Learning prototype via placeholder for zero-shot recognition. *arXiv* **2022**, arXiv:2207.14581.
24. Chen, L.; Zhang, H.-W.; Xiao, J.; Liu, W.; Chang, S. Zero-shot visual recognition using semantics preserving adversarial embedding networks. In Proceedings of the IEEE Conference on Computer Vision and Pattern Recognition, Salt Lake City, UT, USA, 18–22 June 2018; pp. 1043–1052.

25. Akata, Z.; Perronnin, F.; Harchaoui, Z.; Schmid, C. Label-embedding for image classification. *IEEE Trans. Pattern Anal. Mach. Intell.* **2016**, *38*, 1425–1438. [CrossRef]
26. Liu, Y.; Zhou, L.; Bai, X.; Gu, L.; Harada, T.; Zhou, J. Information bottleneck constrained latent bidirectional embedding for zero-shot learning. *arXiv* **2020**, arXiv:2009.07451.
27. Xian, Y.; Lampert, C.H.; Schiele, B.; Akata, Z. Zero-Shot Learning-A Comprehensive Evaluation of the Good, the Bad and the Ugly. *IEEE Trans. Pattern Anal. Mach. Intell.* **2019**, *41*, 9. [CrossRef]
28. Zhao, B.; Wu, B.; Wu, T.; Wang, Y. Zero-shot learning posed as a missing data problem. In Proceedings of the IEEE International Conference on Computer Vision Workshops, Venice, Italy, 22–29 October 2017; pp. 2616–2622.
29. Wang, D.; Li, Y.; Lin, Y.; Zhuang, Y. Relational knowledge transfer for zero-shot learning. In Proceedings of the Thirtieth AAAI Conference on Artificial Intelligence, Phoenix, AZ, USA, 12–17 February 2016; pp. 2145–2151.
30. Changpinyo, S.; Chao, W.L.; Gong, B.; Sha, F. Synthesized classifiers for zero-shot learning. In Proceedings of the IEEE Conference on Computer Vision and Pattern Recognition, Las Vegas, NV, USA, 27–30 June 2016; pp. 5327–5336.
31. Shigeto, Y.; Suzuki, I.; Hara, K.; Shimbo, M.; Matsumoto, Y. Ridge Regression, Hubness, and Zero-shot Learning. In Proceedings of the Machine Learning and Knowledge Discovery in Databases: European Conference, ECML PKDD 2015, Porto, Portugal, 7–11 September 2015; pp. 135–151.
32. Ji, Z.; Yu, X.; Yu, Y.; Pang, Y.; Zhang, Z. Semantic-guided class-imbalance learning model for zero-shot image classification. *IEEE Trans. Cybern.* **2021**, *52*, 6543–6554. [CrossRef]
33. Chen, S.-M.; Wang, W.J.; Xia, B.H.; Peng, Q.M.; You, X.G.; Zheng, F.; Shao, L. Free: Feature re-finement for generalized zero-shot learning. In Proceedings of the IEEE/CVF International Conference on Computer Vision, Montreal, QC, Canada, 10–17 October 2021; pp. 122–131.
34. Li, J.; Jing, M.M.; Lu, K.; Ding, Z.; Zhu, L.; Huang, Z. Leveraging the invariant side of generative zero-shot learning. In Proceedings of the IEEE/CVF Conference on Computer Vision and Pattern Recognition, Long Beach, CA, USA, 16–17 June 2019; pp. 7402–7411.
35. Keshari, R.; Singh, R.; Vatsa, M. Generalized zero-shot learning via over-complete distribution. In Proceedings of the IEEE/CVF Conference on Computer Vision and Pattern Recognition, Seattle, WA, USA, 13–19 June 2020; pp. 13300–13308.
36. Schonfeld, E.; Ebrahimi, S.; Sinha, S.; Darrell, T.; Akata, Z. Generalized zero- and few-shot learning via aligned variational autoencoders. In Proceedings of the IEEE/CVF Conference on Computer Vision and Pattern Recognition, Long Beach, CA, USA, 16–17 June 2019; pp. 8247–8255.
37. Shen, Y.; Qin, J.; Huang, L.; Liu, L.; Zhu, F.; Shao, L. Invertible zero-shot recognition flows. In Proceedings of the European Conference on Computer Vision, 16th European Conference, Glasgow, UK, 23–28 August 2020; pp. 614–631.
38. Yao-Hung, H.T.; Huang, L.-K.; Salakhutdinov, R. Learning robust visual-semantic embeddings. In Proceedings of the IEEE International Conference on Computer Vision, Venice, Italy, 22–29 October 2017; pp. 3591–3600.
39. Yu, Y.; Ji, Z.; Li, X.; Guo, J.; Zhang, Z.; Ling, H.; Wu, F. Transductive zero-shot learning with a self-training dictionary approach. *IEEE Trans. Cybern.* **2018**, *48*, 2908–2919. [CrossRef]
40. Zhu, X.L.; He, Z.L.; Zhao, L.; Dai, Z.C.; Yang, Q.L. A Cascade Attention Based Facial Expression Recognition Network by Fusing Multi-Scale Spatio-Temporal Features. *Sensors* **2022**, *22*, 1350. [CrossRef]
41. Sun, Y.; Bi, F.; Gao, Y.E.; Chen, L.; Feng, S.T. A Multi-Attention UNet for Semantic Segmentation in Remote Sensing Images. *Symmetry* **2022**, *14*, 906. [CrossRef]
42. Liu, R.; Tao, F.; Liu, X.; Na, J.; Leng, H.; Wu, J.; Zhou, T. RAANet: A Residual ASPP with Attention Framework for Semantic Segmentation of High-Resolution Remote Sensing Images. *Remote Sens.* **2022**, *14*, 3109. [CrossRef]
43. Obeso, A.M.; Benois-Pineau, J.; Vazquez, M.S.G.; Acosta, A.Á.R. Visual vs internal attention mechanisms in deep neural networks for image classification and object detection. *Pattern Recognit.* **2022**, *123*, 108411. [CrossRef]

Article

A New Bolt Defect Identification Method Incorporating Attention Mechanism and Wide Residual Networks

Liangshuai Liu *, Jianli Zhao, Ze Chen, Baijie Zhao and Yanpeng Ji

Electric Power Research Institute, State Grid Hebei Electric Power Co., Ltd., Shijiazhuang 050013, China
* Correspondence: liuliangshuai214@163.com; Tel.: +86-186-3390-0355

Abstract: Bolts are important components on transmission lines, and the timely detection and exclusion of their abnormal conditions are imperative to ensure the stable operation of transmission lines. To accurately identify bolt defects, we propose a bolt defect identification method incorporating an attention mechanism and wide residual networks. Firstly, the spatial dimension of the feature map is compressed by the spatial compression network to obtain the global features of the channel dimension and enhance the attention of the network to the vital information in a weighted way. After that, the enhanced feature map is decomposed into two one-dimensional feature vectors by embedding a cooperative attention mechanism to establish long-term dependencies in one spatial direction and preserve precise location information in the other direction. During this process, the prior knowledge of the bolts is utilized to help the network extract critical feature information more accurately, thus improving the accuracy of recognition. The test results show that the bolt recognition accuracy of this method is improved to 94.57% compared with that before embedding the attention mechanism, which verifies the validity of the proposed method.

Keywords: deep learning; bolt defect recognition; wide residuals; double attention

1. Introduction

Bolts are the most numerous and widely distributed fasteners in transmission lines. As they play an important role in maintaining the stable operation of the lines, it is necessary to inspect the abnormal state of the bolts promptly so as to guarantee the safe and steady operation of the lines [1,2]. At present, the use of unmanned aerial vehicles (UAV) equipped with high-resolution cameras for transmission line inspection is not only safer and more efficient [3], but also can integrate deep learning-based image processing technology, which remarkably improves the quality and speed of inspection work. It is of great significance to study the bolted defect image recognition method based on deep learning.

Since the LeNet model was proposed, convolutional neural network models have shown considerable potential in image recognition tasks and have continued to develop. AlexNet [4] further increased the network depth and won the ImageNet challenge in 2012, and then ZFNet [5] and Google Inception Network (GoogLeNet) [6] were proposed one after another. Visual Geometry Group Network (VGGNet) [7] uses 16 convolutional layers and fully connects layers to improve the image recognition accuracy. However, the deepening of the network is not infinite. With the deepening of the number of network layers, problems caused by the deep network such as gradient disappearance and gradient explosion also emerge. The residual network (ResNet) proposed in [8] employs a jump connection method which effectively reduces the parameter number of the network, improves the training speed of the network, and ensures high accuracy. It is an effective solution to the problem that deep neural networks are difficult to train. Based on this, wide residual networks (WRNs) [9] further improve the model performance and increase the recognition accuracy by adding the number and width of convolutional layers to the residual blocks.

Currently, deep learning has been comprehensively used in bolt detection [10], defect classification [11], etc. In [12], the authors used multi-scale features extracted by cascade

Citation: Liu, L.; Zhao, J.; Chen, Z.; Zhao, B.; Ji, Y. A New Bolt Defect Identification Method Incorporating Attention Mechanism and Wide Residual Networks. *Sensors* **2022**, *22*, 7416. https://doi.org/10.3390/s22197416

Academic Editor: Sylvain Girard

Received: 29 August 2022
Accepted: 27 September 2022
Published: 29 September 2022

regions with a convolutional neural network (Cascade R-CNN) to build a path aggregation feature pyramid, which completes bolt defect identification. In [13], the authors enhanced the model complexity and improved the image recognition accuracy through the combined utilization of multiple algorithms. In [14], the authors used wide residuals as the backbone network and selected the optimal structure to achieve effective recognition of bolt defects by adjusting the network-widening dimension. In [15], a bolt defect data augmentation method was proposed based on random pasting, and it effectively expanded the number of bolt defect samples and improved the accuracy of defect recognition. However, due to the small size of the bolt itself, the bolt image features of the aerial transmission line are difficult to extract, and the bolt defect recognition effect is not satisfactory. The above method did not take into account the features of the bolt itself when improving the model.

The attention mechanism can help the network improve the feature extraction ability of the image [16,17]. It is a bionic of human vision that enables the acquisition of detailed information and the suppression of irrelevant information by allocating more attention to the target area. In the domain of deep learning, the attention mechanism uses the feature map to learn a new weight distribution, which is imposed on the original feature map. This weighting not only preserves the original information of the image extracted by the original network, but also enhances focus on the target region, effectively improving the performance of the model. The attention mechanism is not a complete network structure, but a plug-and-play lightweight module. When this module is embedded in the network, it can reasonably allocate computational resources and significantly increase the neural network performance at the cost of a finite increase in the number of parameters. Thus, it has received much attention in detection, segmentation, and recognition tasks because of its practicality and robustness [18–20]. Currently, it can be classified into three categories: spatial domain, channel domain, and hybrid domain. The squeeze and excitation attention network (SENet) [21] and efficient channel attention networks (ECA-Net) [22] are both of single-way attention frames that help the network detect or identify targets better by aggregating information in the spatial domain or channel domain and adaptively learning new weights. These networks are more concise than those with multi-way attention. The selective kernel network (SK-Net) [23] decomposes the feature map into feature vectors by decomposition, aggregation, and matching. In this way, the network is able to extract more detailed feature information. The convolutional block attention module (CBAM) [24] aggregates spatial and channel information to guide the model to focus on the key target regions in the image, while channel attention (CA) improves the ability to capture targets by aggregating one-dimensional channel and spatial information to relate the location relationships between targets in the feature graph. In [25], the authors proposed a dynamic supervised knowledge distillation method for bolt defect recognition and classification by applying knowledge distillation techniques to the bolt defect classification task and combining spatial channel attention. This method effectively improves the accuracy of bolt defect classification. In [26], the authors used an attention mechanism to locate the possible regions of the bolt in the image and then combined it with a deconvolutional network to build a model to achieve accurate detection of the bolt. This is an attention-based mechanism for transmission tower bolt detection. In [27], the authors embedded a dual-attention mechanism in faster regions with a convolutional neural network (Faster R-CNN) to analyze and enhance visual features at different scales and different locations, which effectively improved the bolt detection accuracy.

Although these methods improve the recognition or detection accuracy of bolts to some extent, they are all based on improving the feature expression capability of bolts without improving the model by combining bolt features. In order to identify bolt defects more accurately, by combining the attention mechanisms, we introduce bolt knowledge into the model and study the bolt defect recognition method incorporating dual attention in this paper. WRN is used as the backbone network, and the attention-wide residual network is designed by embedding squeeze and excitation networks [21] and coordinate attention [28] to enhance the network's perception of features in the spatial dimension and channel dimension. The network was designed to enhance its ability to perceive features in the spatial dimension and channel dimension, extracting richer feature information. It is combined with the prior knowledge of bolts to achieve high-accuracy recognition of bolt defects.

2. Materials and Methods

In this work, WRN is used as the backbone network, and the number of channels is 16 × k, 32 × k, and 64 × k, a total of three levels. Among them, three wide residual blocks are in the first level, four wide residual blocks are in the second level, and six wide residual blocks are in the third level. The width factor k is taken as 2. The attention-wide residual network is designed by fusing the attention mechanism in the WRN, so as to enhance the extraction ability for bolt features and improve the accuracy of defect recognition. The overall structure is shown in Figure 1. Firstly, SENet attention is added to each level in the WRN to enhance the network's ability to capture bolt defect features and output higher-quality feature maps. Secondly, CA attention based on structural prior knowledge is imported in combination with the spatial location relationship of pins and nuts on bolts, which enables the network to better utilize the feature location relationship and thus improve the accuracy of bolt defect recognition.

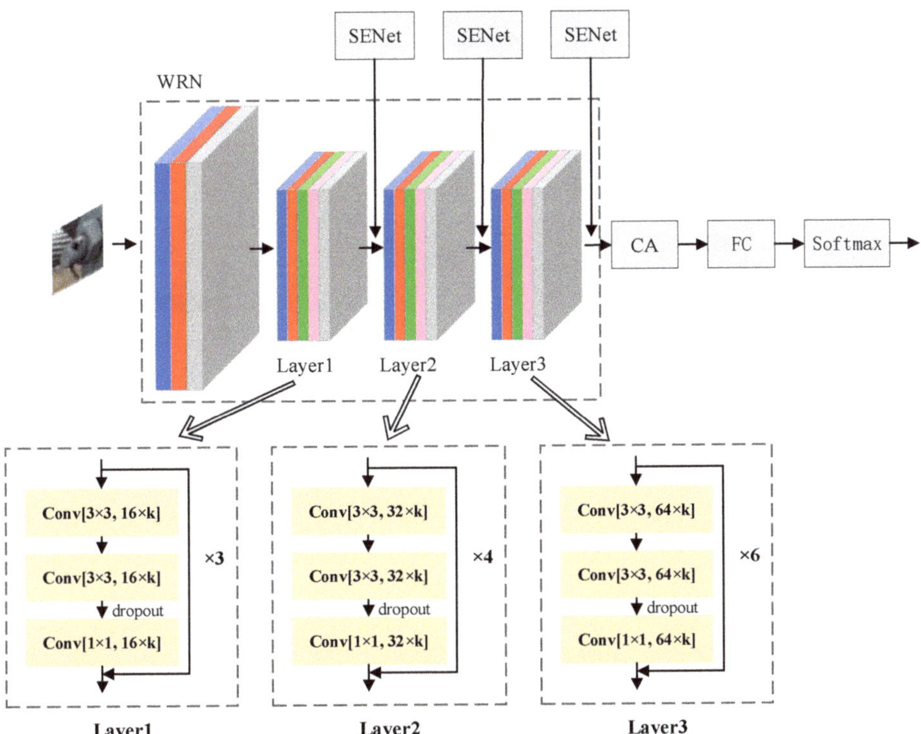

Figure 1. Attention to wide residual network structure.

2.1. WRN Framework for Fusing Channel Attention

A residual network consists of a residual block. It is a constant mapping of shallow features to deeper features using a jump connection so that the residual block can learn more feature information based on the input features and effectively solve the degradation problem caused by deeper networks. However, as the number of network layers increases, the residual block itself cannot be better expressed. A new type of residual approach,

WRN, which widens the number of convolutional kernels in the original residual block, was proposed. It effectively improves the utilization of the residual block, reduces the model parameters, speeds up the computation, and makes it possible to obtain a better training result without a deeper network layer. In addition, WRN adds a dropout between the convolutional layers in the residual block to form a wide ResNet block, which has the effect of improving the performance of the network. The relationship between the ResNet block and the wide ResNet block is shown in Figure 2, where 3×3 indicates the size of the convolution kernel, N is the number of channels, and k indicates the width factor.

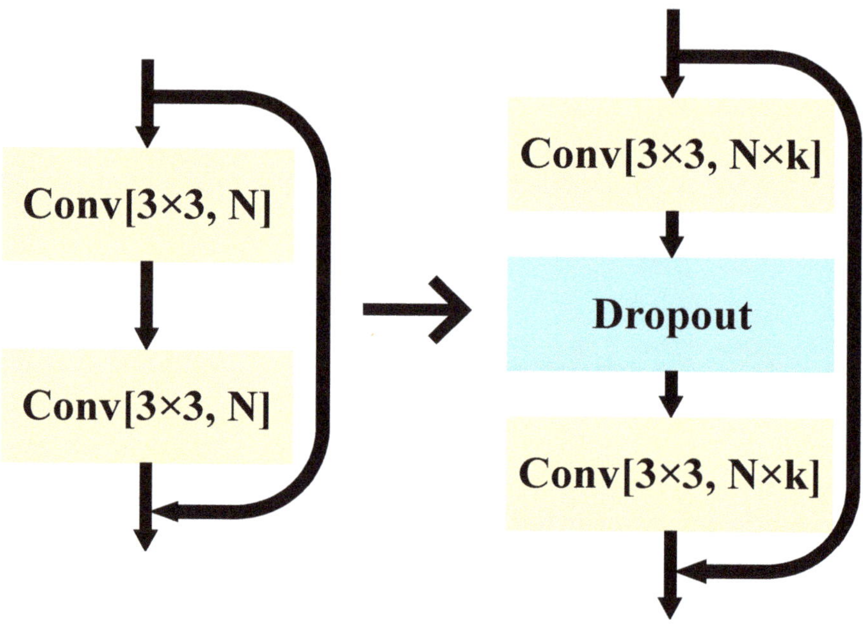

Figure 2. Schematic diagram of the relationship between ResNet block (**left**) and wide-ResNet block (**right**).

SENet attention can aggregate the information from the input features at the spatial level and adaptively acquire new weight relationships through learning. These weight relationships represent the importance of different regions in the feature map, making the network focus on key regions in the feature map as a whole. It helps the information transfer in the network and continuously updates parameters in the direction that is beneficial to the recognition task.

After fusing SENet attention in the WRN, the network first compresses the spatial dimension of the feature map of the input SENet through global average pooling, aggregating spatial information to perceive richer global features of the image and enhancing the network expression capability. The SENet attention structure diagram is shown in Figure 3. The global average pooling operation generates a feature map of $C \times 1 \times 1$ (where C represents the number of channels) to obtain the global information of channels. Then, the correlation between channels is captured by the two fully connected layers with the activation function of ReLu, and the normalized channel weights are then generated by the sigmoid activation function. At this point, the channel weights of dimension $C \times 1 \times 1$ can be multiplied with the input features of dimension $C \times H \times W$ (where H represents the feature map of height, W represents the feature map of width) as a new parameter, i.e., the aligned channel dimension C. For each $H \times W$ matrix, a channel coefficient c is multiplied to obtain the output features $C \times H \times W$ after SENet attention optimization, which enhances the key region features and suppresses irrelevant features to improve the performance of the network.

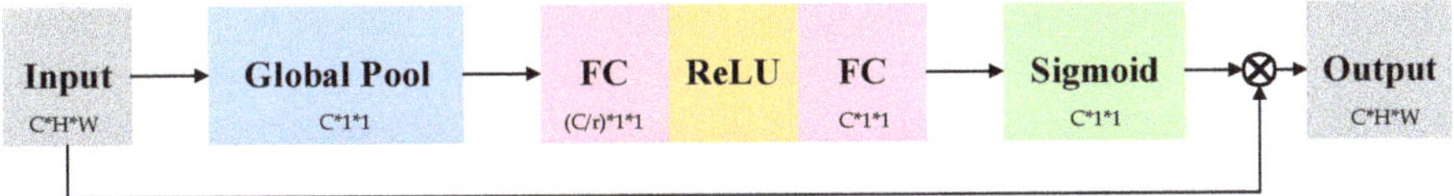

Figure 3. SENet attention structure diagram.

The attention weights are multiplied by the input features to obtain the output features F, as follows:

$$F = \delta(MLP(Pool(F_0))) \times F_0 \tag{1}$$

where F_0 denotes the input features, δ and MLP denote the sigmoid activation function and neural network operation, respectively, and $Pool$ represents the pooling operation.

2.2. CA Attention with Integrated Knowledge

The WRN incorporating SENet attention is enhanced to extract bolt features. However, according to the prior knowledge of the bolt, pins distribute at the head of the bolt while nuts usually locate at the root of the bolt, and these positional relationships are fixed. In order to further improve the bolt defect recognition accuracy using the bolt position information, we add CA attention to the output section of the WRN to enhance the positional relationships of the target. The CA attention structure is shown in Figure 4. First, CA attention decomposes the input features into a horizontal perceptual feature vector of dimension $C \times H \times 1$ and a vertical perceptual feature vector of dimension $C \times 1 \times W$ by global averaging pooling in both directions. The one-dimensional feature vectors in the horizontal and vertical directions are as follows:

$$z_c^h(h) = \frac{1}{W} \sum_{0 \leq i < W} F_c(h, i) \tag{2}$$

$$z_c^w(w) = \frac{1}{H} \sum_{0 \leq j < H} F_c(j, w) \tag{3}$$

where H and W represent the height and width, respectively, h, w, i, and j represent the location coordinates in the feature map, c represents the number of channels, z_c^h represents the one-dimensional feature vector in the horizontal direction, z_c^w represents the one-dimensional feature vector in the vertical direction, and F_c represents the input feature map.

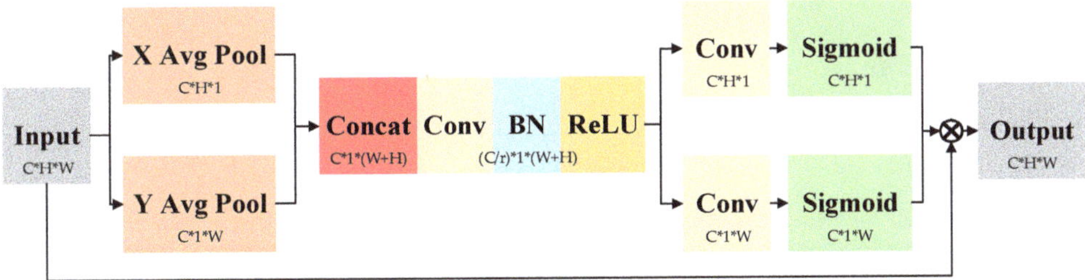

Figure 4. CA attention structure diagram.

In this process, the attention mechanism establishes long-term dependencies in one spatial direction and preserves precise location information in the other, helping the network locate key feature regions more accurately. It also gives the network a better global sensory view of the field as well as rich feature information. Next, the perceptual feature vectors in both directions are aggregated, and the feature mapping is obtained by dimensionality reduction through 1×1 convolution. Unique feature mappings are generated using two one-dimensional features.

$$f = MLP([z^h, z^w]) \tag{4}$$

where $[z^h, z^w]$ represents the stitching operation of two one-dimensional features, and f is the feature mapping of spatial information in the encoding process of horizontal and vertical directions. Finally, the feature mapping is decomposed and normalized by the Sigmoid function to obtain the attention weights in the two directions, and the attention weights in the two directions are multiplied with the input features of dimensionality $C \times H \times W$ to obtain the output features of dimensionality $C \times H \times W$. The two directional weights and output features are as follows:

$$g^h = \delta(T(f^h)) \tag{5}$$

$$g^w = \delta(T(f^w)) \tag{6}$$

$$F(i,j) = F_c(i,j) \times g_c^h(i) \times g_c^w(j) \tag{7}$$

where T represents the convolution operation and $F(i, j)$ is the output feature. After the feature map is processed by CA attention, it is easier for the network to capture the key feature information in the map using location information, and the relationship between channels is more obvious.

3. Test Results and Analysis

3.1. Test Data and Settings

Dataset Construction: We constructed a transmission line bolt defect recognition dataset by cropping and optimizing transmission line aerial images based on the Overhead Transmission Line Defect Classification Rules (for Trial Implementation). Tests were conducted to verify the effectiveness of this method. The dataset was divided into three categories, namely normal bolts, missing pin bolts, and missing nut bolts. There are a total of 6327 images, of which 2990 were normal bolts, 2802 were missing pin bolts, and 535 were missing nut bolts, and the training set and test set were divided in a ratio of 4:1. The samples of each category are shown in Figure 5.

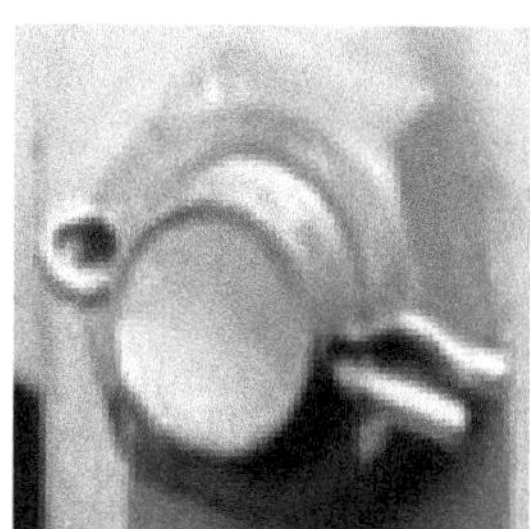
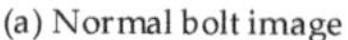

(a) Normal bolt image

(b) Missing pin bolt image

(c) Nut missing bolt image

Figure 5. Three categories of bolt image samples.

Test Settings: The test hardware environment was Linux Ubuntu 16.04, and the GPU used is an NVIDIA GeForce 1080Ti with 11 GB of RAM. The test parameters were a batch size of 64, an epoch count of 200, and a learning rate of 0.1. We used the model to perform a recognition validation on the test set after the model completes an epoch training, obtain and save the accuracy and loss function values of the model on the test set, and take the highest recognition accuracy on the test set as the model evaluation metric after the model completes training. The accuracy rate was chosen as the evaluation index, and the formula is shown in Equation (8), where TP is the number of correctly predicted positive samples, TN is the number of correctly predicted negative samples, FN is the number of incorrectly predicted negative samples, and FP is the number of incorrectly predicted positive samples.

$$\text{Accuracy} = \frac{TP}{TP + TN + FP + FN} \tag{8}$$

3.2. Ablation Tests and Analysis

In order to verify the effectiveness of this method in the actual bolt defect recognition task, we compared the accuracy of the test set under different methods by ablation experiments separately, as shown in Table 1. As can be seen, the recognition accuracy of the base model WRN was 93.31%, an improvement of 0.58% after adding SENet attention. This is because the SENet attention mechanism acquired richer bolt features by compressing spatial information, which enhanced the expressiveness of the network. With the addition of CA attention to the model, the attention mechanism builds long-term dependencies in space and the network is more likely to use the location relationships to capture key feature information, resulting in a 0.72% increment in recognition accuracy. The recognition accuracy of the model was improved by 1.26% after embedding both SENet attention and CA attention. The mutual association between the attentions further improved the network's performance and it has accomplished a more accurate bolt defect recognition task.

Table 1. Ablation test results.

Method	Accuracy (%)
WRN	93.31
WRN + SENet	93.89
WRN + CA	94.03
Ours	94.57

Figure 6 shows the variation curve of the recognition accuracy of the model on the test set as the number of training rounds increases. As can be seen, between epochs of 1 and 60, the accuracy of the model has the fastest rising trend, but the fluctuation is large, and the model has not learned efficient defect recognition ability. Between 60 and 120 epochs, the model's learning task is initially completed, but the accuracy curve is still fluctuating. As the model was trained iteratively, the fluctuation of the accuracy curve gradually decreased after 120 epochs, and finally stabilized after 160 epochs.

Figure 7 shows the loss descent curves of different networks on the training set during the training process. As can be seen, the loss function convergence curves of the model training process under different approaches are compared. The first convergence was between epochs 1 and 60, during which the WRN model had the highest initial value, the WRN plus SENet had the slowest convergence, and the WRN plus CA attention had the fastest convergence. The second convergence was between epochs 60 and 120, and the third was between epochs 120 and 160. In these two convergence domains, the convergence rates and convergence trends of the four models were more or less the same, and the loss function convergence curves of each model showed slight fluctuations. The convergence trend of WRN is the weakest. WRN plus SENet and WRN plus CA attention are similar, and the convergence trend of our proposed method is the best.

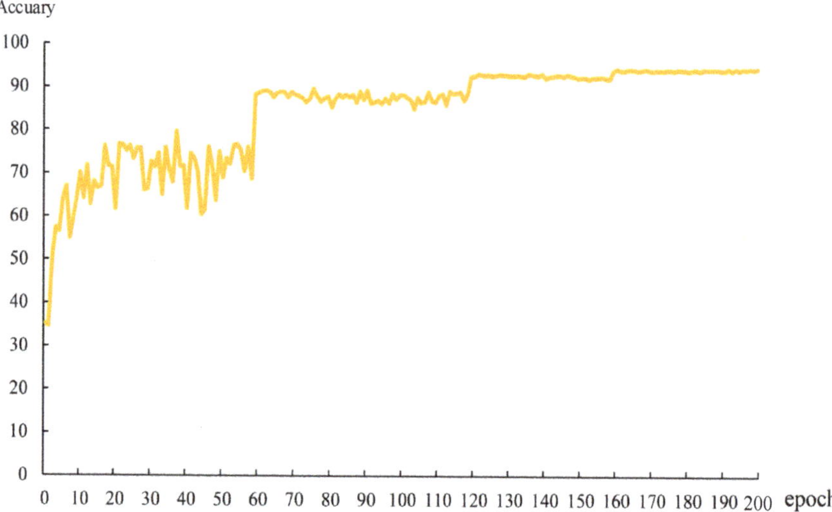

Figure 6. Accuracy curve on test set.

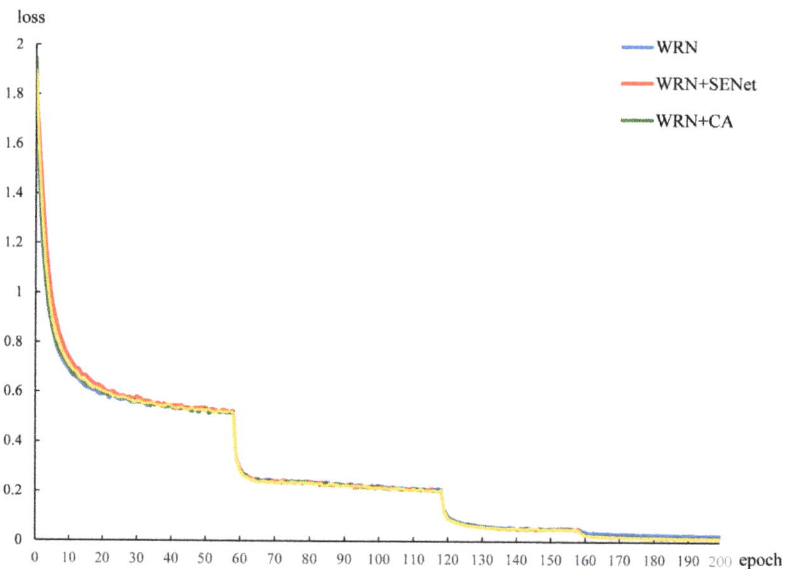

Figure 7. Convergence curve of the model training loss function.

In order to demonstrate the improvement in model performance by attention more intuitively, we used the gradient-weighted class activation mapping (Grad-CAM) [29] algorithm to visualize the feature maps before and after the model improvement, as shown in Figure 8. In this test, a bolt image with missing pins was used as the reference. It can be seen from the figure that the attention area of the features extracted by WRN only is relatively scattered, which is not conducive to the recognition of the bolt by the model. Our method incorporates both SENet attention and CA attention, and the extracted feature map is more significant and discriminative compared with the previous ones. Our method effectively removes redundant information and allows the model to better distinguish bolt categories.

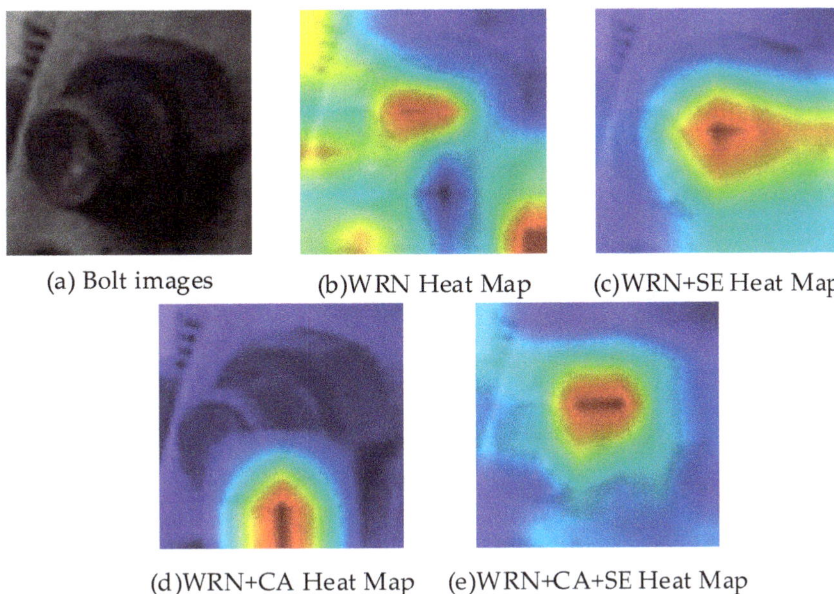

(a) Bolt images (b)WRN Heat Map (c)WRN+SE Heat Map

(d)WRN+CA Heat Map (e)WRN+CA+SE Heat Map

Figure 8. Visualization of the bolt feature map.

3.3. Comparative Tests and Analysis

In these tests, we compared the recognition accuracy of different recognition models for bolt defects in the test set, as shown in Table 2. WRN has the highest accuracy of 93.31%, 3.94% higher than VGG16, and 0.86% and 0.64% higher than ResNet50 and ResNet101, respectively. It fully demonstrates the feasibility and superiority of the backbone network selected in this paper, and paves the way for the next model improvement.

Table 2. Ablation test results.

Recognition Model	Accuracy of Bolt Defect Recognition %
VGG16	89.37
ResNet50	92.45
ResNet101	92.67
WRN	93.31

Meanwhile, we compared the recognition accuracy of each bolt before and after the improvement in the test set, as shown in Figure 9. As can be seen from the figure, after the improvement, the recognition accuracy was increased by 0.77% for normal bolts, 1.24% for missing pin bolts, and 1.76% for missing nut bolts. The accuracy improvement for normal bolts is less, while the accuracy improvement for bolts with missing pins and bolts with missing nuts is more significant with the help of the attention mechanism. This shows that the joint attention-wide residual method proposed in this paper is effective for bolt defect recognition. Embedding SENet attention into each layer to improve the ability of model feature extraction and combining CA attention to focus more accurately on the area of pin or nut in the figure helps the model to better discriminate the bolt category and improve the recognition accuracy.

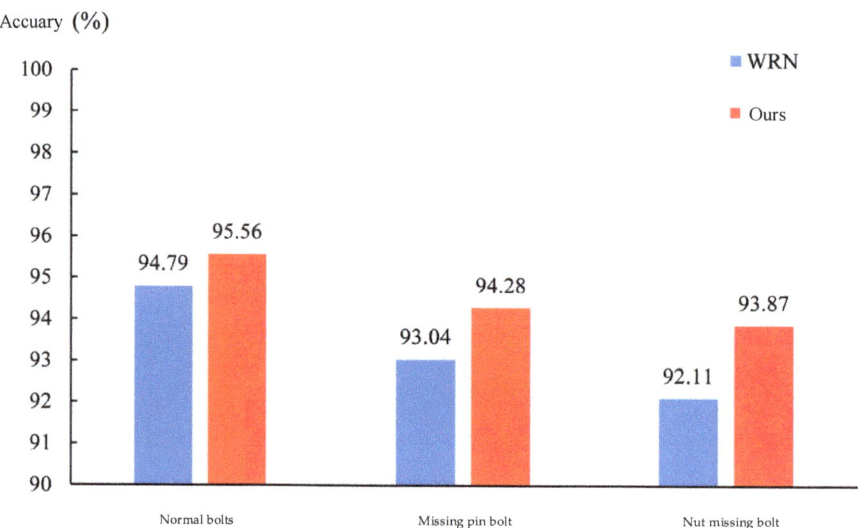

Figure 9. Comparison of classification accuracy before and after model improvement.

4. Conclusions

In order to identify bolt defects more accurately, by taking WRN as the backbone network, we address the problem of difficult extraction of bolt features and the fixed position relationship of pins and nuts on top of the bolts. A new bolt defect identification method incorporating an attention mechanism and wide residual networks is proposed, embedding SENet and CA attention and fusing bolt knowledge. The proposed method can locate the key feature areas with better precision through collaborative space and channel information so as to help the model to improve the recognition accuracy. The proposed method has been validated on a homemade transmission line bolt defect recognition dataset. The test results show that the accuracy of this method was improved by 1.26% compared with that before improvement, which lays a foundation for the transmission line bolt defect detection task.

Author Contributions: Conceptualization, L.L. and J.Z.; methodology, L.L. and J.Z.; software, L.L.; validation, Z.C.; formal analysis, L.L. and B.Z.; investigation, Y.J.; resources, L.L.; data curation, L.L.; writing—original draft preparation, J.Z.; writing—review and editing, L.L.; visualization, Z.C.; supervision, Y.J.; project administration, L.L.; funding acquisition, L.L. All authors have read and agreed to the published version of the manuscript.

Funding: This paper was supported by the State Grid Hebei Electric Power Provincial Company Science and Technology Project Fund Grant Project: Research on visual defect detection technology of power transmission and transformation equipment based on deep knowledge reasoning (project number: kj2021-016).

Institutional Review Board Statement: Not applicable.

Informed Consent Statement: Not applicable.

Data Availability Statement: Not applicable.

Acknowledgments: Special thanks are given to Power grid Technology center, Electric Power Research Institute, State Grid Hebei Electric Power Co., Ltd., Shijiazhuang 050013, China.

Conflicts of Interest: All authors have received research grants from Electric Power Research Institute, State Grid Hebei Electric Power Co., Ltd. None of the authors have received a speaker honorarium from the company or own stock in the company. None of the authors have been involved as consultants or expert witnesses for the company. The content of the manuscript has not been applied for patents; none of the authors are the inventor of a patent related to the manuscript content.

Abbreviations

The following abbreviations are used in this manuscript.

UAV	Unmanned Aerial Vehicle
GoogLeNet	Google Inception Network
VGGNet	Visual Geometry Group Network
ResNet	Residual Network
WRN	Wide Residual Networks
Cascade R-CNN	Cascade Regions with Convolutional Neural Network
SENet	Squeeze and Excitation Attention Network
ECA-Net	Efficient Channel Attention Networks
SK-Net	Selective Kernel Network
CBAM	Convolutional Block Attention Module
CA	Channel Attention
Faster R-CNN	Faster Regions with Convolutional Neural Network
Grad-CAM	Gradient-Weighted Class Activation Mapping

References

1. Zhao, Z.; Qi, H.; Qi, Y.; Zhang, K.; Zhai, Y.; Zhao, W. Detection Method Based on Automatic Visual Shape Clustering for Pin-Missing Defect in Transmission Lines. *IEEE Trans. Instrum. Meas.* **2020**, *69*, 6080–6091. [CrossRef]
2. Han, Y.; Han, J.; Ni, Z.; Wang, W.; Jiang, H. Instance Segmentation of Transmission Line Images Based on an Improved D-SOLO Network. In Proceedings of the 2021 IEEE 3rd International Conference on Power Data Science, Harbin, China, 26 December 2021; pp. 40–46.
3. He, T.; Zeng, Y.; Hu, Z. Research of Multi-Rotor UAVs Detailed Autonomous Inspection Technology of Transmission Lines Based on Route Planning. *IEEE Access* **2019**, *7*, 114955–114965. [CrossRef]
4. Krizhevsky, A.; Sutskever, I.; Hinton, G.E. Imagenet classification with deep convolutional neural networks. In *Advances in Neural Information Processing Systems*; MITP: Boston, MA, USA, 2012; pp. 1097–1105.
5. Zeiler, D.; Fergus, R. Visualizing and understanding convolutional networks. In Proceedings of the European Conference on Computer Vision, Zurich, Switzerlan, 6–12 September 2014; Springer: Cham, Switzerland, 2014; pp. 818–833.
6. Szegedy, C.; Liu, W.; Jia, Y. Going deeper with convolutions. In Proceedings of the IEEE Conference on Computer Vision and Pattern Recognition, Boston, MA, USA, 7–12 June 2015; IEEE: Piscataway, NJ, USA, 2015; pp. 1–9.
7. Simonyan, K.; Zisserman, A. Very deep convolutional networks for large-scale image recognition. In Proceedings of the International Conference on Learning Representations, San Diego, CA, USA, 7–9 May 2015.
8. Xie, S.; Girshick, R.; Dollá, P. Aggregated residual transformations for deep neural networks. In Proceedings of the IEEE Conference on Computer Vision and Pattern Recognition, Honolulu, HI, USA, 21–26 July 2017; IEEE: Piscataway, NJ, USA, 2017; pp. 1492–1500.
9. Zagoruyko, S.; Komodakis, N. Wide residual networks. *arXiv* **2016**, arXiv:1605.07146.
10. Wang, C.; Wang, N.; Ho, S.-C.; Chen, X.; Song, G. Design of a New Vision-Based Method for the Bolts Looseness Detection in Flange Connections. *IEEE Trans. Ind. Electron.* **2020**, *67*, 1366–1375. [CrossRef]
11. Xiao, L.; Wu, B.; Hu, Y. Missing Small Fastener Detection Using Deep Learning. *IEEE Trans. Instrum. Meas.* **2021**, *70*, 1–9. [CrossRef]
12. Wang, H.; Zhai, X.; Chen, Y. Two-stage pin defect detection model based on improved Cascade R-CNN. *Sci. Technol. Eng.* **2021**, *21*, 6373–6379.
13. Zhao, Y.Q.; Rao, Y.; Dong, S.P. Survey on deep learning object detection. *J. Image Graph.* **2020**, *25*, 629–654.
14. Qi, Y.; Jin, C.; Zhao, Z. Optimal Knowledge Transfer Wide Residual Network Transmission Line Bolt Defect Image Classification. *Chin. J. Image Graph.* **2021**, *26*, 2571–2581.
15. Zhao, W.; Jia, M.; Zhang, H.; Xu, M. Small Target Paste Randomly Data Augmentation Method Based on a Pin-losing Bolt Data Set. In Proceedings of the 2021 IEEE 3rd International Conference on Power Data Science, Harbin, China, 26 December 2021; pp. 81–84.
16. Brauwers, G.; Frasincar, F. A General Survey on Attention Mechanisms in Deep Learning. *IEEE Trans. Knowl. Data Eng.* **2021**. [CrossRef]

17. Sun, J.; Jiang, J.; Liu, Y. An Introductory Survey on Attention Mechanisms in Computer Vision Problems. In Proceedings of the 2020 6th International Conference on Big Data and Information Analytics (BigDIA), Shenzhen, China, 4–6 December 2020; pp. 295–300.

18. Li, Y.-L.; Wang, S. HAR-Net: Joint Learning of Hybrid Attention for Single-Stage Object Detection. *IEEE Trans. Image Process.* **2020**, *29*, 3092–3103. [CrossRef] [PubMed]

19. Guo, Z.; Huang, Y.; Wei, H.; Zhang, C.; Zhao, B.; Shao, Z. DALaneNet: A Dual Attention Instance Segmentation Network for Real-Time Lane Detection. *IEEE Sens. J.* **2021**, *21*, 21730–21739. [CrossRef]

20. Lian, S.; Jiang, W.; Hu, H. Attention-Aligned Network for Person Re-Identification. *IEEE Trans. Circuits Syst. Video Technol.* **2021**, *31*, 3140–3153. [CrossRef]

21. Hu, J.; Shen, L.; Albanie, S. Squeeze-and-Excitation Networks. *IEEE Trans. Pattern Anal. Mach. Intell.* **2020**, *42*, 2011–2023. [CrossRef] [PubMed]

22. Wang, Q.; Wu, B.; Zhu, P. ECA-Net: Efficient channel attention for deep convolutional neural networks. In Proceedings of the IEEE Conference on Computer Vision and Pattern Recognition, Seattle, WA, USA, 13–19 June 2020; IEEE: Piscataway, NJ, USA, 2020; pp. 11531–11539.

23. Li, X.; Wang, W.; Hu, X. Selective kernel networks. In Proceedings of the IEEE Conference on Computer Vision and Pattern Recognition, Long Beach, CA, USA, 16–20 June 2019; IEEE: Piscataway, NJ, USA, 2019; pp. 510–519.

24. Woo, S.; Park, J.; Lee, J.Y. CBAM: Convolutional block attention module. In Proceedings of the Computer Vision-ECCV 2018, Munich, Germany, 8–14 September 2018; Springer International Publishing: Cham, Switzerland, 2018; pp. 3–19.

25. Zhao, Z.; Jin, C.; Qi, Y. Image Classification of Bolt Defects in Transmission Lines Based on Dynamic Supervised Knowledge Distillation. *High Volt. Technol.* **2021**, *47*, 406–414.

26. Weitao, L.; Huimin, G.; Qian, Z.; Gang, W.; Jian, T.; Meishuang, D. Research on Intelligent Cognition Method of Missing status of Pins Based on attention mechanism. In Proceedings of the 2021 IEEE 4th Advanced Information Management, Communicates Electronic and Automation Control Conference, Chongqing, China, 18–20 June 2021; IEEE: Piscataway, NJ, USA, 2021; pp. 1923–1927.

27. Qi, Y.; Wu, X.; Zhao, Z.; Shi, B.; Nie, L. Faster R-CNN Aerial Photographic Transmission Line Bolt Defect Detection Embedded with Dual Attention Mechanism. *Chin. J. Image Graph.* **2021**, *26*, 2594–2604.

28. Hou, Q.B.; Zhou, D.Q.; Feng, J.S. Coordinate Attention for Efficient Mobile Network Design. In Proceedings of the IEEE Conference on Computer Vision and Pattern Recognition, Nashville, TN, USA, 20–25 June 2021; IEEE: Piscataway, NJ, USA, 2021; pp. 13708–13717.

29. Selvaraju, R.R.; Cogswell, M.; Das, A.; Vedantam, R.; Parikh, D.; Batra, D. Grad-cam: Visual explanations from deep networks via gradient-based localization. In Proceedings of the IEEE International Conference on Computer Vision, Venice, Italy, 22–29 October 2017; pp. 618–626.

 *sensors*

Article

Multi-Patch Hierarchical Transmission Channel Image Dehazing Network Based on Dual Attention Level Feature Fusion

Wenjiao Zai [†] and Lisha Yan *,[†]

College of Engineering, Sichuan Normal University, Chengdu 610101, China; zaiwenjiao@sicnu.edu.cn
* Correspondence: yanlisha@stu.sicnu.edu.cn; Tel.: +86-187-8384-2873
[†] These authors contributed equally to this work.

Abstract: Unmanned Aerial Vehicle (UAV) inspection of transmission channels in mountainous areas is susceptible to non-homogeneous fog, such as up-slope fog and advection fog, which causes crucial portions of transmission lines or towers to become fuzzy or even wholly concealed. This paper presents a Dual Attention Level Feature Fusion Multi-Patch Hierarchical Network (DAMPHN) for single image defogging to address the bad quality of cross-level feature fusion in Fast Deep Multi-Patch Hierarchical Networks (FDMPHN). Compared with FDMPHN before improvement, the Peak Signal-to-Noise Ratio (PSNR) and Structural Similarity Index Measure (SSIM) of DAMPHN are increased by 0.3 dB and 0.011 on average, and the Average Processing Time (APT) of a single picture is shortened by 11%. Additionally, compared with the other three excellent defogging methods, the PSNR and SSIM values DAMPHN are increased by 1.75 dB and 0.022 on average. Then, to mimic non-homogeneous fog, we combine the single picture depth information with 3D Berlin noise to create the UAV-HAZE dataset, which is used in the field of UAV power assessment. The experiment demonstrates that DAMPHN offers excellent defogging results and is competitive in no-reference and full-reference assessment indices.

Keywords: transmission channels; non-homogeneous fog; dual attention; DAMPHN; image defogging

Citation: Zai, W.; Yan, L. Multi-Patch Hierarchical Transmission Channel Image Dehazing Network Based on Dual Attention Level Feature Fusion. *Sensors* **2023**, *23*, 7026. https://doi.org/10.3390/s23167026

Academic Editors: Ke Zhang and Yincheng Qi

Received: 17 July 2023
Revised: 4 August 2023
Accepted: 4 August 2023
Published: 8 August 2023

1. Introduction

UAVs have been increasingly employed in power inspection to find safety problems effectively [1]. However, in hilly regions, advection fog, uphill fog, and valley fog are frequently encountered [2,3], causing critical portions of transmission lines or towers to become fuzzy or even wholly concealed and decreasing fault detection accuracy. Image-defogging technology can be used to address the appeal issues. However, the non-homogenous fog is challenging for the current homogenous fog removal method. Additionally, the initial non-homogeneous defogging method FDMPHM exploits residual connections between several levels and ignores the issues with channel redundancy and unequal pixel distribution in cross-level fusion. Based on this, we suggest the Dual Attention Level Feature Fusion Multi-Patch Hierarchical Network (DAMPHN), which aims to enhance the cross-level fusion method of FDMPHN and produce superior defogging effects. Haze non-uniformity is not considered in power inspection image defogging studies due to a lack of non-homogeneous haze datasets. Therefore, to create a dataset that may represent non-homogeneous haze in mountainous places (UAV-HAZE), this paper ingeniously combines image depth measurements with 3D Berlin noise. The suggested DAMPHN performs better in color preservation and haze removal than the other four advanced approaches and can complete the picture preprocessing of transmission channels, according to numerous experiments on three open datasets and UAV-HAZE.

1.1. Related Work

Model-based parameter estimation and model-free picture enhancement methods are currently the main single-image fog removal categories. Additionally, future images for machine vision services will be of higher quality because of advancements in CCD image technology [4]. Some researchers have used the image defog technique to preprocess photos based on high-quality photographs for transmission channels.

1.1.1. Model-Based Parameter Estimation Method

By predicting the transmission matrix $t(x)$ and global atmospheric light A from the haze graph $J(x, \lambda)$, these approaches, based on the atmospheric scattering model [5] provide images $I(x, \lambda)$ that are devoid of haze. In Equation (1), the atmospheric scattering model is displayed.

$$I(x, \lambda) = t(x)J(x, \lambda) + A(1 - t(x)) \tag{1}$$

$$t(x) = e^{-\beta(\lambda)d(x)} \tag{2}$$

where $d(x)$ denotes the depth of the scene and $\beta(\lambda)$ the scattering coefficient. Both the early dark channel prior (DCP) [6] and the color decay prior (CAP) [7] were put out and offered concepts for further study. Convolutional neural networks (CNN) were later developed, and Cai et al. [8] used CNNs with various kernel parameters for the first time to extract the distinctive information of dark channel, color attenuation, maximum contrast, and hue disparity to solve the parameters. Li et al. [9] equalized $t(x)$ and A as a parameter based on Formula (1) and applied CNN and residual connection to get this parameter. Zhang et al. [10] used the Dense-Net and U-net networks, respectively. A Densely Connected Pyramid Dehazing Network (DCPDN) was subsequently proposed based on the joint discriminator of adversarial networks and the optimization parameter estimate of the edge retention loss function. To achieve adaptive fusion, Li et al. [11] employed a multi-stage deep convolutional network to estimate $t(x)$ and A and added a memory network and a two-level attention mechanism to determine the weight of findings at each stage. To filter haze residuals step by step and achieve dehazing, Li et al. [12] modified Formula (1) to be task-oriented and assembled recurrent neural networks based on encoder-decoder and space. Bai et al. [13], who combined $t(x)$ and A into a single parameter and calculated it using the depth pre-defamer. The progressive feature fusion module and the picture recovery module were created to improve parameter estimation.

1.1.2. Model-Free Image Enhancement Method

This technique uses a coding-decoding structure to directly learn the link between the haze/clear image mapping and integrates attention mechanisms, feature fusion, and other techniques to enhance the dehazing performance. Das et al. [14] introduced the Fast Deep Multi-Patch Hierarchical Network (FDMPHN) and Fast Multi-Scale Hierarchical Network (FDMSHN) by improving the loss function, which was inspired by literature [15]. According to Wang et al. [16], a heterogeneous twin network was suggested, U-Net was used to extract haze features, and a detail enhancer network was set up to improve image details. Liu et al. [17] proposed an attention-based multi-scale defogging network (GridDehazeNet), which introduced a channel attention mechanism to improve feature fusion ability among multiple scales. A feature fusion attention network with a channel and pixel focus that prioritizes high-frequency and dense hazy areas was proposed by Qin et al. [18]. To improve the ability to extract edge texture features, Wang et al. [19] created the edge branch module based on the multi-level attention dehazing module and the feature fusion module based on Laplace gradient prior knowledge. Using extended convolution in the multi-scale part, channel attention mechanism in the cross-level fusion part, and frequency domain loss in the loss function part, Yang et al.'s [20] combination of FDMPHN and FDMSHN methods to obtain dense feature maps produced good results. A transfer attention technique was created by Wang et al. [21] to deal with non-uniform noise in images. To focus on the non-uniform hazy region and address the issues of artifacts and excessive smoothing, Zhao et al. [22] developed a dynamic attention module based on the dual attention mechanism. Guo et al. [23] suggested a self-paced half-course learning-driven attention image-generating technique based on the dual attention mechanism to enhance the ability to clear regions with considerable brightness disparities of fog.

1.1.3. Transmission Channel Image Dehazing Method

Recently, researchers have used it in power inspection after taking inspiration from the appeal algorithm. Liu et al. [24] created their own UAV picture collection for transmission line inspection and used the DCPDN approach to achieve dehazing. To address the drawbacks of the DCP method, Zhang et al. [25] divided the sky region by fusing the Canny operator and gradient energy function to obtain a more accurate atmospheric light value, and Zhai et al. [26] optimized the quadtree segmentation method. Both techniques were then applied to the image dehazing of transmission line monitoring systems. To remove haze from photographs of an insulator umbrella disk in transmission lines, Xin et al. [27] coupled a limited-contrast adaptive histogram equalization method with the dark channel, bright channel, and these methods. Gao et al.'s [28] use of DCP to remove haze from fixed-point monitoring photographs of a tower or pole was likewise based on this technique. Yan et al. [29] created their dataset for UAV power inspection and used FDMPHN to achieve dehazing.

1.2. Motivation and Contribution

The model-based parameter estimate methods produces improved outcomes in the area of picture fog removal. However, the overall image that DCP restored is dark, and color distortion can easily happen in areas of bright light. The reduction impact is weak when the depth of field shift in the image is not visible or when there is haze, as CAP is dependent on the color saturation of the image. To maximize the fog removal effect, later researchers used CNN to estimate the parameters $t(x)$ and A. However, both the parameter estimation methods based on CNN [8,10,11] and the parameter estimation method after the improved atmospheric scattering model [9,12,13] are subject to artifacts, color distortion, and haze residues because of the shortcomings of the atmospheric scattering model. Although the model-free image enhancement methods are not limited by the model, it depends on the ability of the network to extract and fuse the haze features. Only residual connections are used in the multi-patch network FDMPHN for cross-level feature fusion, disregarding channel differences and pixel distribution non-uniformity. Therefore, when the non-uniform characteristics of haze or the fog area are strong, it is easy for haze residue and detail blur to appear. Later researchers enhanced the network's capacity for feature extraction by improving the attention mechanism [17–23], but it was also challenging to address the issue of non-uniform fog.

In the area of fog removal in power inspection images, Refs. [24–28] all use a uniform haze dataset created based on an atmospheric scattering model as the foundation for their analyses, neglecting the non-uniform characteristics of haze distribution in natural settings. As a result, it is only appropriate for processing images with uniform haze distribution. It performs poorly when dealing with powerful light sources and non-uniform haze, and the image quality after recovery is also subpar. Furthermore, power inspection picture fog removal is still in the uniform haze removal stage, and it is challenging to make progress due to the relative paucity of non-uniform haze datasets [30]. Therefore, this paper suggests a Dual Attention Level Feature Fusion Multi-Patch Hierarchical Network (DAMPHN) to enhance the defogging effect of UAV inspection photos of transmission lines in mountainous terrain. This work's key contributions can be summed up as follows:

1. It is suggested to use a Dual Attention Level Feature Fusion Multi-Patch Hierarchical Network (DAMPHN) that combines an encoder-decoder module with a Dual Attention Level Feature Fusion (DA) module. The experimental results show that the network has low color distortion and a good defogging effect.
2. DA module is proposed. DA makes use of channel attention, pixel attention, and residual connection to enhance the multi-patch layered network's cross-level feature function strategy. The DA module has strong feature fusion capabilities, as demonstrated by numerous ablation tests.

3. By calculating picture depth information and inserting 3D Berlin noise of various frequencies, 2225 pairs of non-homogeneous haze/clear images datasets are constructed based on the actual situation. The dataset can, as closely as possible, mimic the characteristics of haze dispersal in mountainous regions. Later, it is employed to support DAMPHN training and testing, which can enhance the ability of UAV inspection photos of transmission lines in mountainous locations to remove fog.

Figure 1 illustrates the specifics of our implementation strategy for DAMPHN-based image preprocessing of mountain areas' transmission channel images. Based on this, Section 2 details the DAMPHN network structure. It also includes the encoder-decoder and DA module's unique construction and the loss function needed for network training. The datasets required for the ablation and application experiments and the creation of the training parameters are described in Section 3. The usefulness of the suggested DA and DAMPHN is first demonstrated in Section 4 through several ablation experiments, after which many algorithms are trained and tested using real haze photos of mountain power transmission routes and UAV-HAZE datasets. Section 5 discusses and analyzes the experimental results. In Section 6, several conclusions are made.

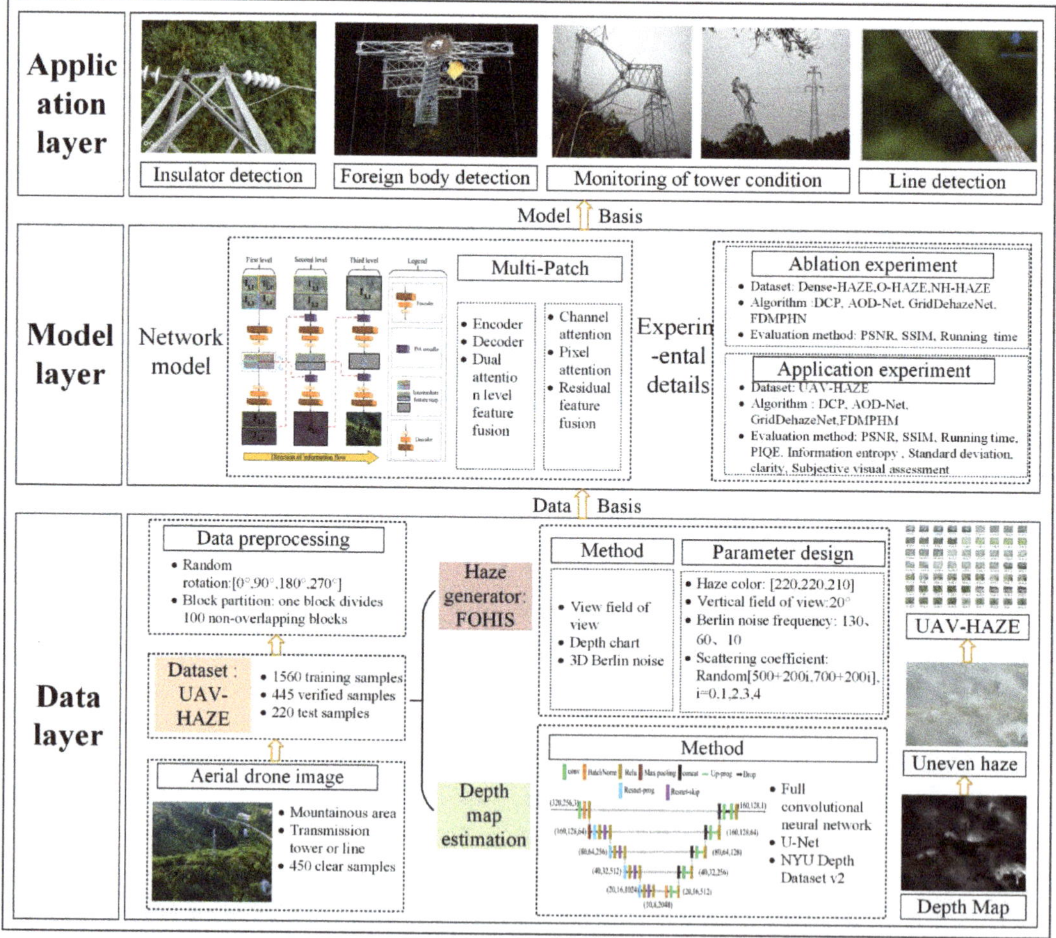

Figure 1. Implementation scheme of image preprocessing of mountain transmission channel based on DAMPHN.

2. Materials and Methods

In this study, the encoder-decoder and DA module-based DAMPHN are suggested. This section's first paragraph introduces DAMPHN's architecture and design principles, as well as those of its submodules. The training and optimization of the DAMPHN loss function are covered in the second section.

2.1. DAMPHN

DAMPHN network is a multi-level structure, and each level comprises corresponding encoders and decoders. The potential of hierarchical feature fusion is further enhanced by a Dual Attention Level Feature Fusion module (DA). Figure 2 displays the structure in its entirety. Figure 2 depicts DAMPHN with i hierarchical structure, where each level processes 4, 2, and 1 picture blocks, respectively, and when $i = 1, 2, 3$. The j block of level i is represented as $I_{i,j}$ if the input image is I. The first layer then divides I into 4 blocks, identified as $I_{1,1}$, $I_{1,2}$, $I_{1,3}$, and $I_{1,4}$, both vertically and horizontally. I is divided vertically into two blocks, designated as $I_{2,1}$ and $I_{2,2}$, by the second stratum. I is directly inputted into the third layer, which is represented as $I_{3,1}$.

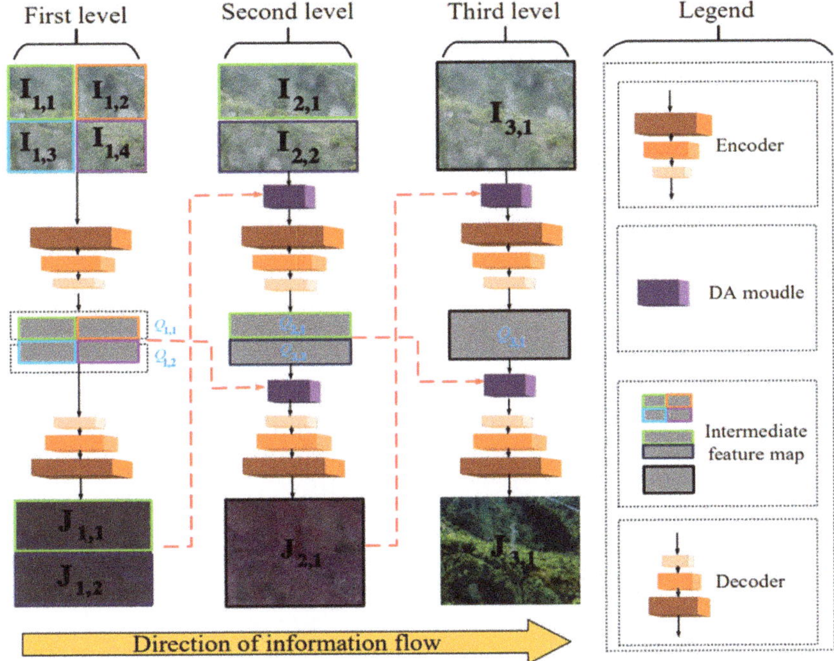

Figure 2. DAMPHN network structure.

The pair of encoder decoders that make up each level are denoted as Enc_i and Dec_i, respectively. The encoding feature $Q_{i,j}$ can be retrieved after the input picture $I_{i,j}$ has sequentially been through the encoder and DA module. In particular, see Equation (3).

$$Q_{i,j} = \begin{cases} Cat\left[Enc_i\left(I_{i,2j-1}\right), Enc_i\left(I_{i,2j}\right)\right], i = 1, j\epsilon 1, 2 \\ Enc_i\left(DA\left(I_{i,j}, J_{i-1,j}\right)\right), i = 2, j\epsilon 1, 2 \\ Enc_i\left(DA\left(I_{i,j}, J_{i-1,j}\right)\right), i = 3, j = 1 \end{cases} \tag{3}$$

The local feature output $J_{i,j}$ of all levels can be acquired after the DA module and decoder. $J_{3,1}$ represents the final dehazing image after DAMPHN feature extraction from the local to the overall concept. The specifics are presented in Equation (4):

$$J_{i,j} = \begin{cases} Dec_i(Q_{i,j}), i = 1, j\epsilon1,2 \\ Dec_i(DA(Cat[Q_{i,j}, Q_{i,2j}], Cat[Q_{i-1,j}, Q_{i-1,2j}])), i = 2, j = 1 \\ Dec_i(DA(Q_{i,j}, Cat[Q_{i-1,j}, Q_{i-1,2j}])), i = 3, j = 1 \end{cases} \tag{4}$$

2.1.1. Encoder-Decoder

The encoder is used to extract the feature data from the image, while the decoder reconstructs the image using the feature data. Three convolution layers and three residual modules (Resblock × 3) make up the encoder in this study. The decoder has a similar design to the encoder, with three residual modules, two transposed convolution layers, and one convolution layer. In order to generate a haze-free image and restore the image scale, decoder transposition convolution is utilized. Figure 3 depicts its network structure.

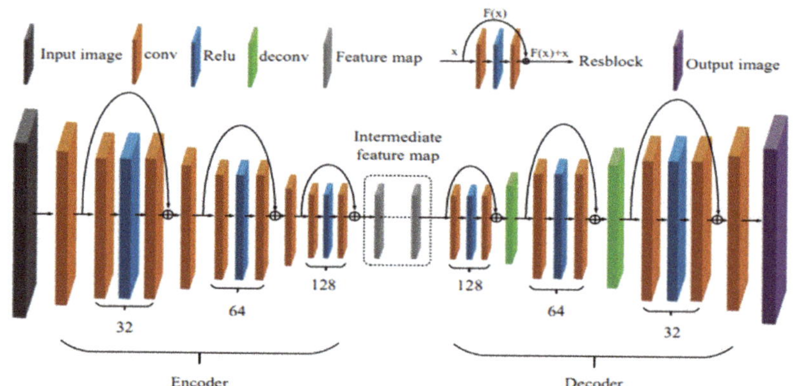

Figure 3. Encoder-decoder module structure.

2.1.2. DA Module

After going through the encoder-decoder during the hierarchical fusion process, the local feature $J_{i,j}$ is produced from the foggy picture I input at the first and second levels. The convolution transformation of $Q_{i,j}$ yields each channel of $J_{i,j}$. As a result, the residual connection in the original FDMPHN network is employed directly in cross-level fusion, and the uneven and redundant channel direction in the fusion feature process is not considered. Additionally, the residual splicing method does not consider the uneven distribution of picture pixels, and the encode-decoder in the original FDMPHN network relies on pixel domain mapping to understand the intricate relationship between the hazy image and the clear image. This led to the development of the DA module provided in this paper, as seen in Figure 4.

The channel domain feature response is first collected by adding the channel attention layer, and subpar or duplicated features are suppressed. Second, by including a pixel attention layer to concentrate on regions of the image with uneven pixel distribution, we may enhance the fusion process' attention to dense haze or high-frequency regions. After stitching, input the channel attention layer (Ca_layer) and pixel attention layer (Pa_layer), assuming that the feature picture of the current level is $F_C\epsilon R^{H\times W\times C}$ and the feature picture of the previous level is $F_U\epsilon R^{H\times W\times C}$. $F_{CA}\epsilon R^{H\times W\times C}$ and $F_{PA}\epsilon R^{H\times W\times C}$ are obtained. Finally, this paper obtains the output F of the final DA module using the convolution joint processing channel and the outcomes of pixel attention processing to make up for the information lost in the extraction process of dual attention layers.

$$F_{CA} = Ca\_layer(Cat[F_C, F_U]) \tag{5}$$

$$F_{PA} = Pa\_layer(F_{CA}) \tag{6}$$

$$F = Cat[conv(F_{PA}), F_{PA}, F_{CA}, Cat[F_C, F_U]] \tag{7}$$

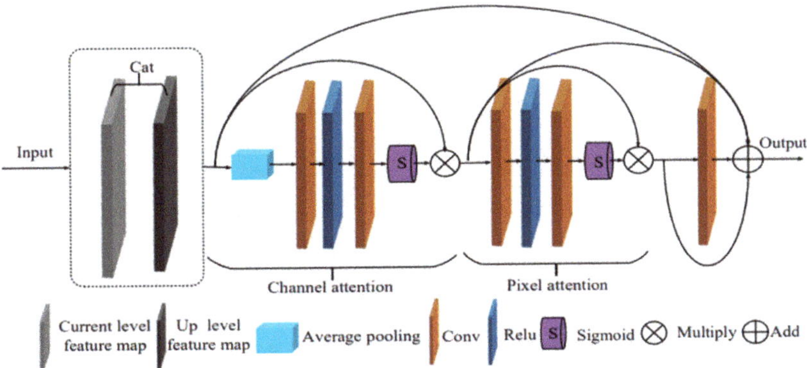

Figure 4. DA module structure.

2.2. Loss of DAMPHN

The total loss function L of DAMPHN is shown in Equation (8), where, respectively, L_r, L_p, and L_{tv} stand for reconstruction loss, perception loss, and total variational loss.

$$L = \alpha_r L_r + \alpha_p L_p + \alpha_{tv} L_{tv} \tag{8}$$

- Reconstruction loss L_r;

Determine the difference between the clear pictures J pixel and the N DAMPHN defogging images J_n. MAE and MSE are combined linearly. L_r can be written as:

$$L_r = \alpha_{r1} \frac{1}{N} \sum_{i=1}^{N} \|J_n - J\| + \alpha_{r2} \frac{1}{N} \sum_{i=1}^{N} \|J_n - J\|^2 \tag{9}$$

- Perception loss L_p;

The VGG16 network was used to calculate features using the pre-trained model. The network's convolution layers (Conv1-2, Conv2-2, and Conv3-2) were utilized to calculate differences, designated as $\varphi(\cdot)$, and extract features. L_p is written as:

$$L_p = \frac{1}{C_K W_K H_K} \sum_{K=1}^{3} \|\varphi_K(J_n) - \varphi_K(J)\| \tag{10}$$

- Total variation loss L_{tv}.

L_{tv} is calculated by computing the gradient amplitude of the dehazing image to reduce noise and keep the image smooth. $\nabla_x(\cdot)$ and $\nabla_y(\cdot)$ in Equation (11), respectively, are used to obtain the gradient matrix of the picture in the horizontal and vertical directions.

$$L_{tv} = \|\nabla_x(J_n)\|_2 + \|\nabla_y(J)\|_2 \tag{11}$$

3. Experiment Setup

3.1. Dataset

3.1.1. Ablation Experimental Dataset

The datasets for the ablation experiment were chosen from three standard datasets from the IEEE CVRP NTIRE Seminar: Dense-HAZE [31], O-HAZE [32], and NH-HAZE [33].

Dense-HAZE includes 55 identical pairs of dense haze/clear images. From the sample, 1–45 pairings were chosen for training, 46–50 pairs for verification, and 51–55 pairs for testing in this study. O-HAZE includes 45 sets of outdoor, non-homogeneous haze/clear images. From that set, 1–35 pairs were chosen for training, 36–40 pairs for verification, and 41–45 pairings for testing in this study. Fifty-five non-homogeneous haze/clear image pairs are included in NH-HAZE. In this study, 1–45 were selected for training, 46–50 for verification, and 51–55 for testing.

3.1.2. Self-Built Transmission Channel Inspection Dataset (UAV-HAZE)

In haze image imaging, because it is often manifested as loss of image visibility, the atmospheric extinction coefficient σ can solve the $\beta(\lambda)$ in Equation (12).

$$\beta(\lambda) = \frac{3.912}{\sigma} \tag{12}$$

Additionally, visibility varies depending on height. Therefore, the depth value of the scene and the vertical field of view of the camera are used to estimate the elevation values of the pixels and their distribution characteristics are calculated to replicate the distribution and color characteristics of genuine haze. To imitate the color features of haze, Formula (1) includes the haze color value I_{al} as follows:

$$I(x, \lambda) = t(x)J(x, \lambda) + A(1 - t(x)) \times I_{al} \tag{13}$$

Taking into account the mountain haze's irregularly distributed properties. Non-uniform haze is created using 3D Berlin noise, and a haze generator called FOHIS [34] is suggested. They are used to mimic non-uniform haze by making three Berlin noises of varying amplitudes and frequencies, which are then merged with Equation (13) and multiplied by $\beta(\lambda)$.

$$P\_noise = \frac{1}{3} \sum_{i=1}^{3} \frac{P\_noise_i}{2^i - 1} \tag{14}$$

In light of FOHIS, this work estimated the picture depth value in order to synthesize the mountain transmission into the UAV-HAZE dataset [35]. In the synthesis process, the I_{al} of the three-color channels of the image RGB is set to [220,220,210], respectively, to simulate the color characteristics of the blue-white mountain fog. Then, to imitate the distribution features of mountain haze, the vertical field of view of the camera is adjusted to $20°$. This is combined with the depth value of picture pixels, and the pixel elevation value is calculated. The non-uniform properties of mountain haze were then simulated by creating 3D Berlin noise with three distinct frequency values (f = 130, 60, 10). Finally, the data [700–900], [900–1100], [1100–1300] and [1300,1500] were chosen as the extinction coefficients in Equation (12) using 450 mountain transmission channel photos obtained by UAV inspection as the original dataset. A total of 2225 non-uniform simulated haze/clear images of various concentrations make up UAV-HAZE, which is divided into training sets, verification sets, and test sets in a ratio of 7:2:1. There are 1560 pairs in the training set, 445 pairs in the verification set, and 220 teams in the test set.

3.2. Implementation Details

NVIDIA GeForce RTX3090 (24 GB) was the platform used for the experiment. Data preprocessing involves cropping each training image into 100 non-overlapping image blocks with a size of 120 × 160 pixels and unifying the image resolution of the training set across Dense-HAZE, O-HAZE, NH-HAZY, and UAV-HAZE to 1200 × 1600. The image blocks were simultaneously rotated at random angles of 0, 90, 180, and 270 degrees. The Adam optimizer is initially employed in DAMPHN network training with exponential decay rates $\gamma_1 = 0.9$, $\gamma_2 = 0.999$, starting learning rates 1×10^{-4}, and batch sizes 100. We also adjusted the learning rate using an equally spaced strategy with step size = 10 and gamma = 0.1. Then, the hyperparameters of the loss function are set to $\alpha_r = 1$, $\alpha_p = 6 \times 10^{-3}$,

$\alpha_{tv} = 2 \times 10^{-8}$, $\alpha_{r1} = 0.6$, $\alpha_{r2} = 0.4$. Finally, when the verification set loss function is stable, the training is stopped and the best model is obtained.

4. Experiment Results

4.1. Ablation Experiment

Two phases of the ablation experiment were conducted. The first and second sections, respectively, confirm the reliability of the DA module and the DAMPHN network.

4.1.1. DA Module

Due to the low cross-level fusion quality of the original multi-patch algorithm FDM-PHN, the DA module is proposed in this study. In order to reduce the complexity of the algorithm, the encoder-decoder structure of FDMPHN is diminished. The three sets of experiments listed below are explicitly included in this section:

(I)   The network encoder-decoder has six residual modules (Resblock $\times$ 6) using only FDMPHN.
(II)   The approach suggested in this work builds on (I) by adding a DA module (FDMPHN + DA). A DA module plus six residual modules (Resblock $\times$ 6) make up the network encoder-decoder.
(III) To optimize (II) and DAMPHN, the solution presented in this research uses just three residual modules (Resblock $\times$ 3).

- Quantitative evaluation

PSNR [36], SSIM [37], and APT were chosen for quantitative evaluation in this section of the experiment. The visual noise and distortion decrease as the PSNR value rises. The recovery of structural properties such as image brightness and contrast is measured by SSIM. The dehazing is better the higher the value. Table 1 displays the precise outcomes of the three groups of studies. In Table 1, when (I) and (II) are compared, the addition of the DA module raised PSNR and SSIM in the three datasets by an average of 0.35 dB and 0.0073, whereas APT rose by 19% (0.007 s). Comparing (I) and (III), the average PSNR and SSIM in the three datasets are raised by 0.30 dB and 0.011, respectively, and APT is shortened by 11% (0.003 s), respectively, after the encode-decoder structure is streamlined.

Table 1. Results of DA module ablation experiments.

	Method	Dense-HAZE			O-HAZE			NH-HAZE		
		PSNR	SSIM	APT	PSNR	SSIM	APT	PSNR	SSIM	APT
(I)	FDMPHN	13.47	0.4369	0.031	19.93	0.7045	0.030	16.87	0.5512	0.030
(II)	FDMPHN + DA	14.03	0.4512	0.036	20.35	0.6976	0.035	16.94	0.5656	0.037
(III)	DAMPHN	13.89	0.4497	0.027	20.20	0.7138	0.027	17.07	0.5621	0.027

- Convergence analysis

This section assessed the convergence using the dynamic curves for training loss, PSNR, and SSIM. On Dense-HAZE, O-HAZE, and NH-HAZE, Figure 5 displays the training losses, PSNR, and SSIM for the FDMPHN, FDMPHN+DA, and DAMPHN approaches, respectively. Figure 5 shows the training and testing of the three approaches on three separate datasets, with the training losses, PSNR, and SSIM information displayed in the rows and columns, respectively. Figure 5a illustrates how the training loss for the aforementioned approaches steadily lowers as the number of iterations increases and gradually stabilizes at 35–40 rounds. In Figure 5b,c, all three approaches converge after 200 rounds, and the DA module performs better regardless of how complicated or straightforward the encoder-decoder structure is.

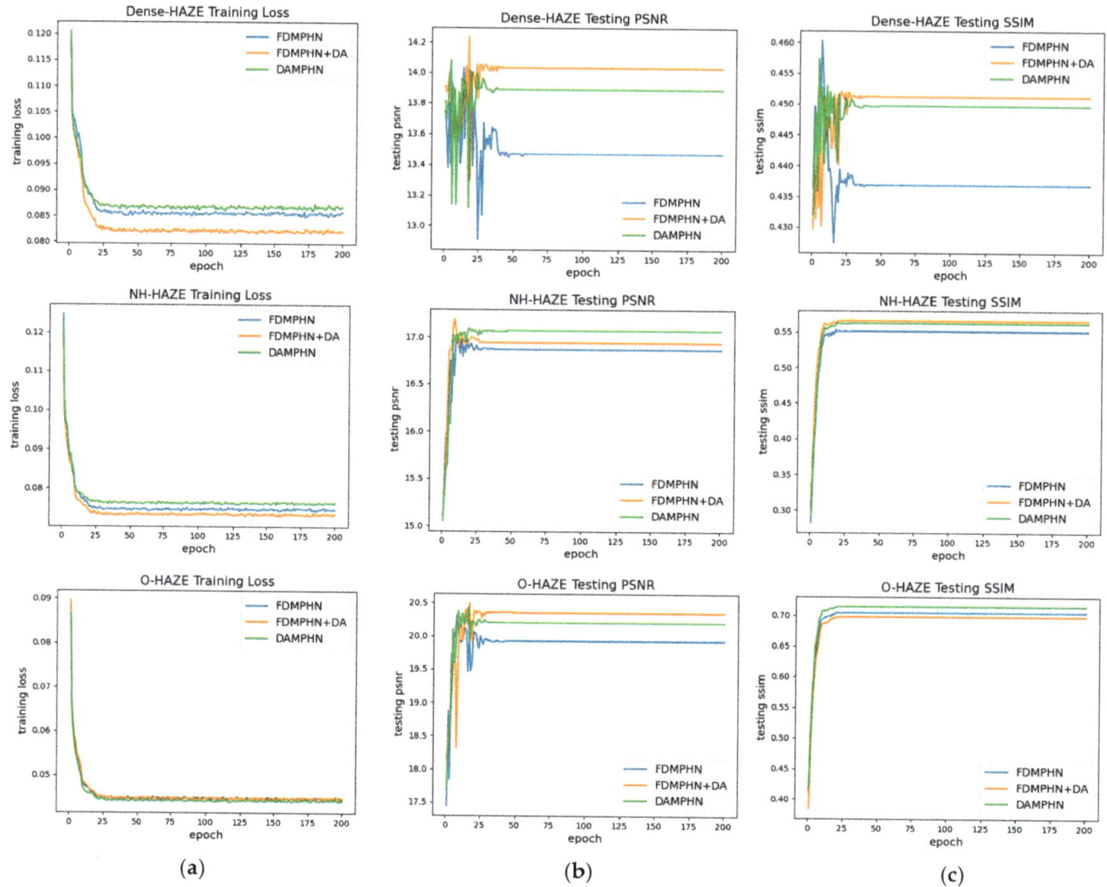

Figure 5. Training loss curve and test PSNR and SSIM curve. (**a**) Training loss. (**b**) Testing PSNR. (**c**) Testing SSIM.

4.1.2. DAMPHN Network

To more accurately evaluate DAMPHN, we further conducted quantitative, qualitative, and convergence evaluation on three datasets, Dense-HAZE, O-HAZE, and NH-HAZE, with DCP [6], AOD-Net [9], FDMPHN [14], and GridDehazeNet [17], respectively.

- Quantitative evaluation

PSNR, SSIM, and APT are also used to gauge how well various techniques remove haze. The outcomes of the quantitative comparison are displayed in Table 2. In Table 2, the blue values represent the optimal values, and the underlined values represent the sub-optimal values. In the three datasets, the PSNR and SSIM values of DAMPHN are 3.72 dB and 0.0666 higher than those of DCP on average, and ART is 94% shorter. The defog quality of AOD-Net in the Dense-HAZE dataset is comparable to that of DAMPHN. However, on the non-uniform haze datasets O-HAZE and NH-HAZE, the PSNR and SSIM values of DAMPHN are increased by 1.72 dB and 0.0446 compared with the average value of AOD-Net. The effect of GridDehazeNet on the fog removal in the three datasets has its own advantages compared with the method in this paper. Specifically, DAMPHN is, on average, 0.38 dB higher than GridDehazeNet's PSNR value, but the SSIM value is lower than GridDehazeNet's 0.025. Finally, compared with FDMPHN in the three datasets, the

PSNR and SSIM values of DAMPHN are increased by 0.30 dB and 0.011 on average, and ART is shortened by 11%.

Table 2. Results of DAMPHN Network quantitative comparison.

Method	Dense-HAZE			O-HAZE			NH-HAZE		
	PSNR	SSIM	APT	PSNR	SSIM	APT	PSNR	SSIM	APT
DCP [6]	11.60	0.3854	0.406	15.66	0.6753	0.440	13.28	0.4650	0.416
AOD-Net [9]	13.85	0.4714	0.023	18.19	0.6950	0.010	15.64	0.4918	0.009
GridDehazeNet [17]	13.50	0.4721	0.026	19.82	0.7108	0.026	16.70	0.6101	0.026
FDMPHN [14]	13.47	0.4369	0.031	19.93	0.7045	0.030	16.87	0.5512	0.030
DAMPHN (ours)	13.89	0.4497	0.027	20.20	0.7138	0.027	17.07	0.5621	0.027

- Qualitative assessment

The experiment's visual comparison component is the main focus here. Among the images, the haze distribution in the first and second rows is more uniform, and the haze distribution in the third and fourth rows is uneven. The DCP results in Figure 6 reveal color distortion and a significant degree of residual haze. The image's color changes to dark yellow after AOD-Net fog removal, and a significant quantity of haze residue remains in the non-uniform haze area. GridDehazeNet has a good fog effect when the haze distribution is relatively uniform, but the image's color after fog removal is darker than that of the clear picture. In addition, in the case of non-uniform haze, GridDehazeNet also shows many haze residues. The image's overall color after fog removal by FDMPHN is closer to the clear image when the haze distribution is more uniform. Still, the color distortion appears on the ground of the first line of the picture. Regarding non-uniform haze, FDMPHN has a good de-fogging effect, but its de-noising solid ability also causes image smoothing, resulting in blurred details. DAMPHN is visually similar to FDMPHN. However, in the enlarged area of the fourth row of the image, the DAMPHN haze residue is less.

- Convergence analysis

In this experiment section, the convergence is assessed using the change curves of PSNR and SSIM with the number of training rounds. Figure 7 shows the results of each round of PSNR and SSIM tests for four de-fogging techniques on three datasets. DCP has the fastest convergence rate. AOD-Net uses a relatively lightweight CNN structure in the parameter estimation process, which has poor stability and the slowest convergence rate. When the PSNR value of the current verification set is assumed to be greater than the previous results during GridDehazeNet training, the round model is optimal. Under dynamic control, its convergence rate ranks fourth. The FDMPHN and DAMPHN set the hyperparameters before training, and the validation set is used to optimize the hyperparameter settings. Therefore, both FDMPHN and DAMPHN converge faster. Specifically, in Figure 7a, DAMPHN converges faster than FDMPHN. In Figure 7b,c, FDMPHN and DAMPHN converge at similar speeds. Therefore, DAMPHN in this paper is in second place in terms of convergence speed.

4.2. Transmission Channel Image

4.2.1. Synthetic Dataset UAV-HAZE

DAMPHN can be utilized to clear haze from Sichuan's mountainous areas' transmission channel scenery. This section is based on the dataset created in Section 3.1.2, UAV-HAZE. With this collection of data, DCP [6], AOD-Net [9], FDMPHN [14], GridDehazeNet [17], and DAMPHN, the approach in this article, are each examined in turn. This section evaluates both the algorithm's quantitative and qualitative performance.

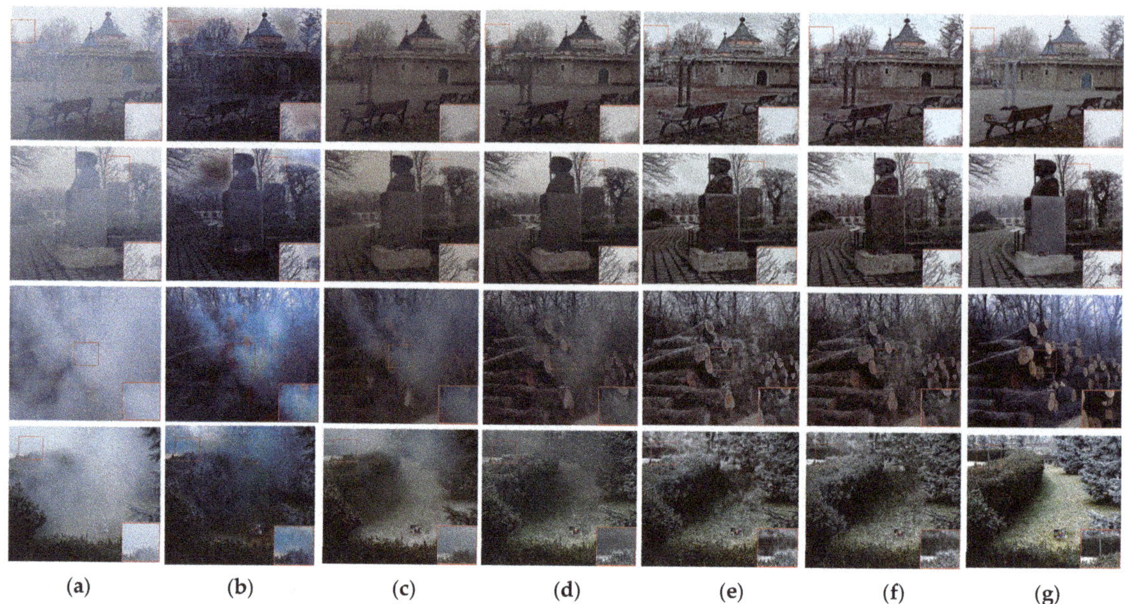

Figure 6. NH-HAZE and O-HAZE dehazing results. (**a**) Hazy. (**b**) DCP. (**c**) AOD-Net. (**d**) GridDe-hazeNet. (**e**) FDMPHN. (**f**) DAMPHN. (**g**) Ground truth.

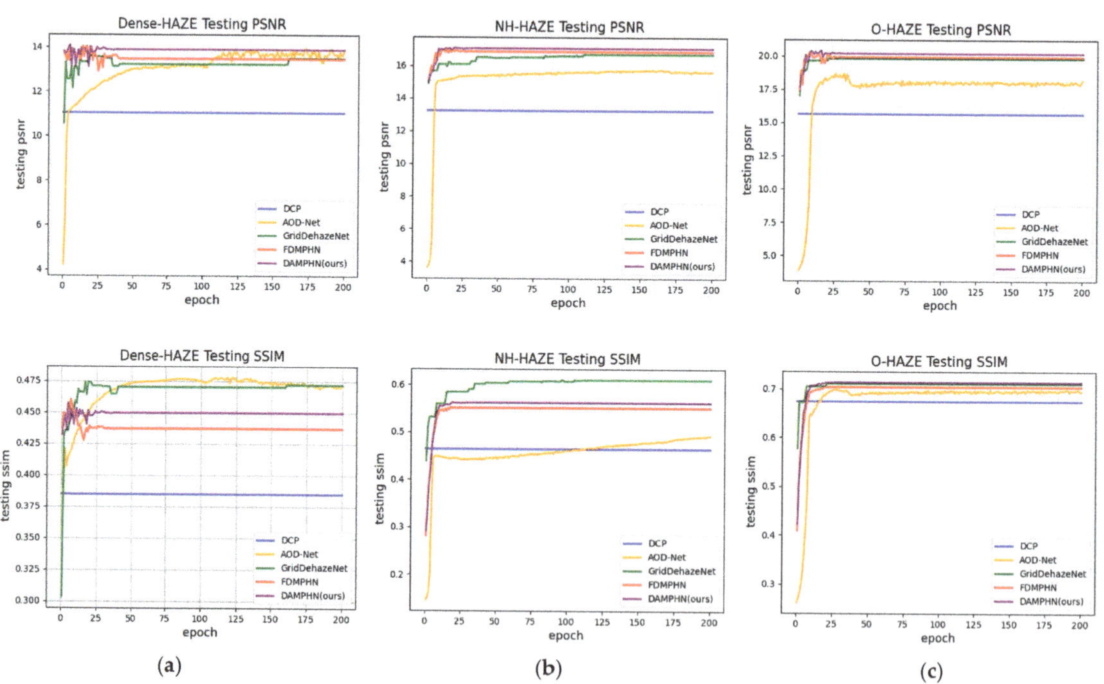

Figure 7. PSNR and SSIM test curves. (**a**) Dense-HAZE. (**b**) NH-HAZE. (**c**) O-HAZE.

- Quantitative evaluation

PSNR, SSIM, and APT were chosen as evaluation indicators. Table 3 presents the experimental outcomes. In Table 3, the blue font is the optimal value, and the underlined value is the sub-optimal value. The PSNR of DAMPHN is optimal, SSIM and ART are sub-optimal. In this study, DAMPHN's PSNR and SSIM values are 7.26 dB and 0.0588 greater than DCP's, respectively. APT barely makes up 4% of DCP techniques. PSNR and SSIM are 9.32 dB and 0.2057 greater in DAMPHN than in AOD-Net, although APT is 14 times higher. The PSNR value of DAMPHN is 0.26 dB higher, and the SSIM value is 0.0007 dB lower than GridDehazeNet. DAMPHN's SSIM value is the same as FDMPHN's, but its PSNR is 0.04 dB higher, and its APT is 94% shorter.

Table 3. Quantitative comparison results on UAV-HAZE.

Method	PSNR	SSIM	APT
DCP [6]	19.97	0.8851	0.352
AOD-Net [9]	17.92	0.7382	0.001
GridDehazeNet [17]	26.98	0.9476	0.015
FDMPHN [14]	27.20	0.9439	0.234
DAMPHN (ours)	27.24	0.9439	0.014

- Qualitative assessment

Figure 8 displays the outcomes of the qualitative comparison between DAMPHN and the techniques mentioned above. DCP has a positive impact in the mist area, according to the analysis of Figure 8. The color of the third row seems distorted when the haze density is excellent, or the randomness of its distribution features is substantial. When dealing with non-uniform haze, AOD-Net's primary result is that a significant amount of haze is left in the processed image, the details are blurred, and there is evident color distortion. The fog removal quality of GridDehazeNet is superior to that of the first two techniques. However, some fog was still present close to the first row's wires and the fourth row's poles and towers. In this study, the FDMPHN and DAMPHN techniques can recover the picture tower's detailed information with excellent clarity and superb color fidelity. FDMPHN does, however, have a trace amount of haze residue in the first row's wire area.

4.2.2. Real Image

The actual utility of DAMPHN was confirmed by the refit project from Gangu to Erlang Mountain in Shuzhou and the real hazy photographs of the Sichuan-Tibet network project. The approach was evaluated using both quantitative and qualitative methodologies.

- Quantitative evaluation

Five non-reference image quality evaluation indexes, including information entropy, standard deviation, clarity, perception-based image quality evaluation method (PIQE) [38], and APT, were chosen for quantitative evaluation because there were insufficient clear reference examples. The more relevant information an image carries, the higher its information entropy. The image's standard deviation is used to assess its contrast; the lower the standard deviation, the more stable the image is. The greater the value, the higher the sharpness, which is defined as the variance of calculating the absolute value of Laplace. Block effects, blur, and noise distortion are calculated using PIQE, and a lower value corresponds to less distortion. In Table 4, the experimental findings are displayed.

In Table 4, the underlined value and the blue text represent the ideal and sub-optimal values, respectively. This approach performs the best regarding clarity and PIQE, comes in second for ART, and comes in third for information entropy and standard deviation. This approach has reduced standard deviation and higher assessment indices compared to DCP. The proposed method has a clear benefit over AOD-Net regarding image quality, but it takes four times as long to operate. DAMPHN has higher evaluation indexes than

GridDehazeNet, except for lower information entropy. DAMPHN is superior to FDMPHN in various assessment indices compared to FDMPHN before improvement, except for the picture information entropy, which is less than 0.17.

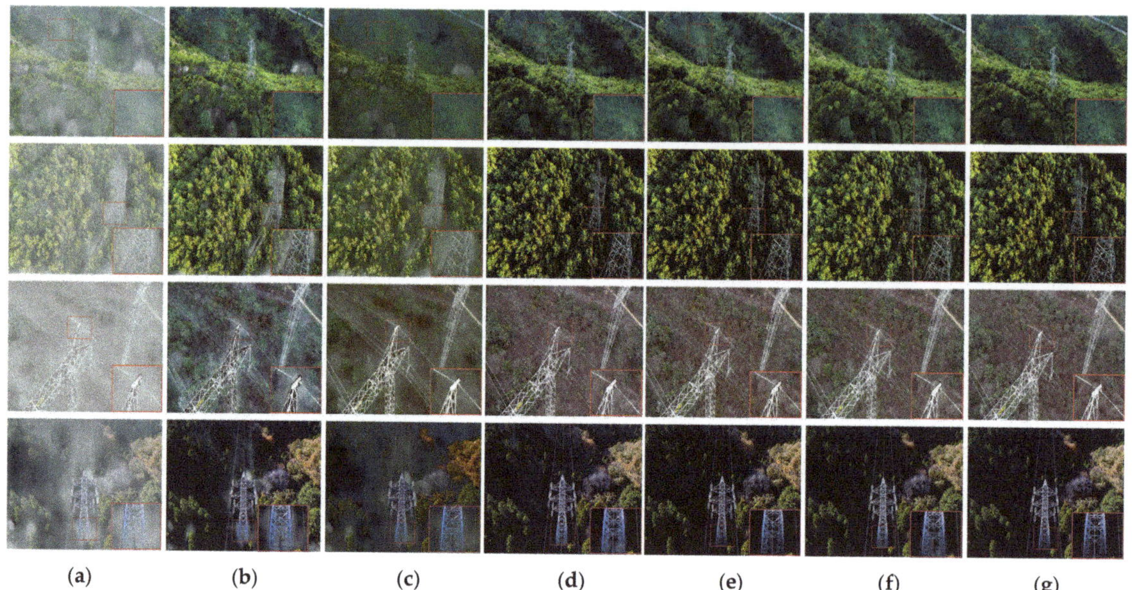

(a) (b) (c) (d) (e) (f) (g)

Figure 8. Results of UAV-HAZE dehazing. (**a**) Hazy. (**b**) DCP. (**c**) AOD-Net. (**d**) GridDehazeNet. (**e**) FDMPHN. (**f**) DAMPHN. (**g**) Ground truth.

Table 4. Results of quantitative evaluation of real images.

Method	Information Entropy	Standard Deviation	Clarity	PIQE	APT
DCP [6]	17.78	32.19	459.86	27.51	0.342
AOD-Net [9]	16.10	45.93	452.79	28.87	0.005
GridDehazeNet [17]	18.28	41.61	470.18	24.90	0.021
FDMPHN [14]	18.10	42.38	465.21	24.48	0.270
DAMPHN (ours)	17.93	41.92	536.11	23.98	0.020

- Qualitative assessment

Figure 9 displays two transmission channel views of the retrofitting project from Gangu to Erlang Mountain in Shuzhou and the haze reduction effect of four groups of the Sichuan-Tibet interconnection project. Uphill fog, uphill fog, advection fog, and radiation fog are all depicted in lines 1 through 4. Intuitive examination reveals that the color of DCP is severely altered and turns blue-purple in the sky area. AOD-Net effectively removes haze. However, it has glaring issues with blurred details and intensified hue. Although GridDehazeNet effectively removes fog, there is still some fog in the third-row valley and second-row tower areas. The image is also slightly lavender once the fog has been eliminated, for instance, the first row's valley fog area and the fourth row's pole tower area. In places with high haze density, such as the tower area in the second row and the valley area in the third row, FDMPHN has a competitive dehazing impact but leaves haze residue behind. This technique also results in color distortion, as seen in how the first row of trees on an ascent turned yellow. After adding a DA module, DAMPHN may now pay closer attention to areas with dense fog and a non-uniform haze. As a result, the method suggested in this paper removes fog more thoroughly than GridDehazeNet and FDMPHN in the first-row and third-row valley areas. Additionally, there is no purple or yellowing in terms of color preservation.

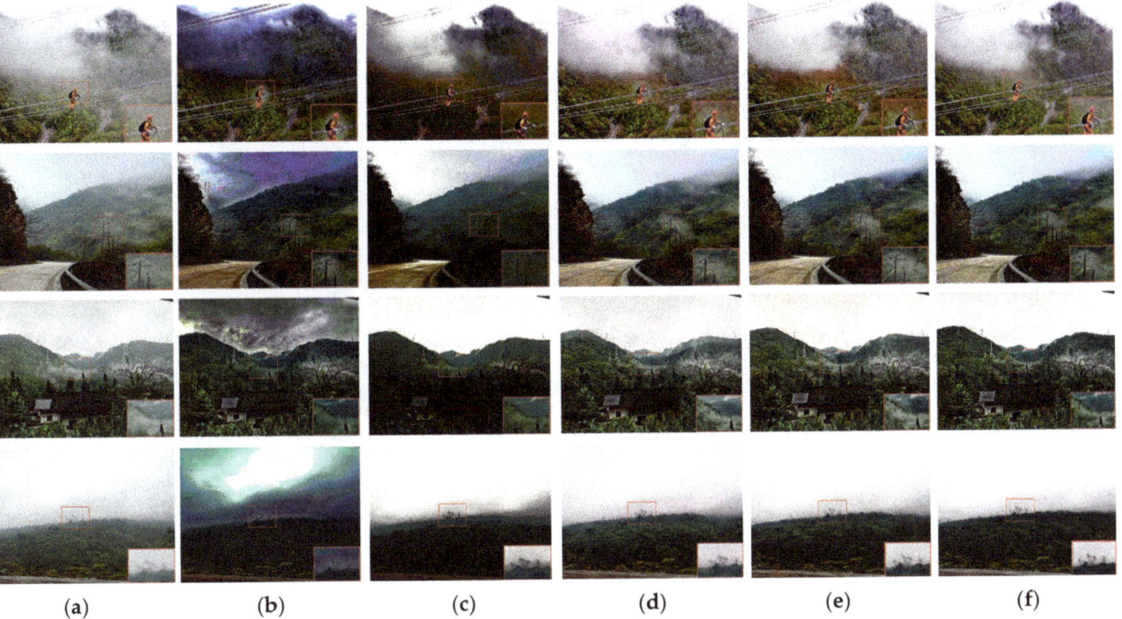

Figure 9. Dehazing result of real transmission channel image. (**a**) Hazy. (**b**) DCP. (**c**) AOD-Net. (**d**) GridDehazeNet. (**e**) FDMPHN. (**f**) DAMPHN.

5. Discussion

In this paper, the issue of transmission line haze that is unevenly dispersed in mountainous places was studied. A DAMPHN is introduced, an innovative non-uniform haze-defogging network model put forth in this research to facilitate picture preprocessing for UAV transmission channel inspection in mountainous terrain. Similarly, the DAMPHN network model is universal. DAMPHN can be used for preprocessing other images in fog environments, such as unmanned visual perception, surveillance video (road traffic, transmission lines), and tachographs. DCP, AOD-Net, GridDenzeNet, and FDMPHN were utilized in numerous tests using open datasets (Dense-HAZE, O-HAZE, and NH-HAZE) and self-built datasets (UAV-HAZE) to demonstrate the efficacy of DAMPHN.

Notably, because the assumption of uniform distribution of air concentration in the atmospheric scattering model limits both DCP and AOD-Net, the error of estimating parameters is significant in dense fog and non-homogeneous haze. DAMPHN is a multi-level end-to-end fog removal network that seeks to remove fog by discovering the relationship between the haze and clear image mapping. DAMPHN does not, therefore, need to estimate the parameters; instead, it relies on the dataset's basis, and the higher the base, the higher the quality of fog removal. GridDehazeNet solves the problem of feature fusion between different scales in multi-scale networks by introducing channel attention. DAMPHN solves the problem of feature fusion between different levels in multi-patch networks by introducing channel and pixel attention mechanisms. GridDehazeNet has vital artifact removal, so the SSIM value is stronger than DAMPHN. DAMPHN pays attention to the problem of uneven pixel distribution, pays attention to the removal of non-uniform fog, and has a strong denoising ability and high PSNR value. FDMPHN is identical to a multi-patch defogging network, but the residual connections in hierarchical fusion restrict how well it can fuse features. The pixel attention layer of the DAMPHN's DA module is designed to pay attention to areas with unequal haze distribution. In contrast, the channel attention layer is designed to appropriately evaluate the channel domain properties. DAMPHN has a better defogging impact as a result than FDMPHN.

Additionally, the frequently used image segmentation algorithms U-Net and GridNet have produced effective outcomes in image segmentation and picture defogging via innovation. DCPDN solves parameter A using the U-Net network. GridDehazeNet proposes a multi-scale attention network based on GridNet. They both have superior defogging effects. With dual U-Net, Amyar et al. [39] created a multi-task and multi-scale network structure that was effectively used for lung tumor segmentation, classification, and prediction. However, DAMPHN accomplishes picture fog removal from the local to the global by helping the feature extraction of the bigger patch image from the top layer with the detailed feature of the lower layer. From the overall to the local picture segmentation, image fog removal, and other tasks, U-Net will employ the more comprehensive information collected from the bottom layer to aid in the development of smaller receptive field information. Consequently, the two networks' designs have produced successful outcomes in their respective domains.

In conclusion, the DAMPHN approach offers an excellent defogging effect, less color distortion, and quick processing speed. In a location with a lot of fog, it is impossible to eliminate it entirely, and the details are hazy. DAMPHN can improve the defog effect by enhancing the encoder-decoder structure, feature extraction, and reconstruction skills, all of which were influenced by U-Net in the field of image segmentation, or by combining with the conventional image edge previous knowledge to increase the texture information and boost the fog removal effect.

6. Conclusions

This paper proposes that DAMPHN can achieve a good defog effect and restore the color and brightness of the image. The network encoder-decoder module and DA module are composed. The former can learn the mapping relationship between haze and clear pictures and has a strong feature extraction ability. The latter enhances the feature fusion ability by empowering the combination of channel attention and pixel attention. However, in excessive haze density, it cannot be entirely removed, and the details are hazy. Future work will improve the haze removal effect by enhancing texture information through edge prior and enhancing the encoder-decoder structure. Additionally, using 3D Berlin noise and image depth information to simulate haze's non-uniform distribution characteristics is not only just restricted to UAV mountain transmission channel inspection; it can also be applied to a broader range of situations to enhance generalization performance.

Author Contributions: Conceptualization, methodology, and writing—review and editing, W.Z.; software, validation, data curation, and writing—original draft preparation, L.Y. All authors have read and agreed to the published version of the manuscript.

Funding: This paper was supported by the following project: Innovation and entrepreneurship training program for college students of China: "Vision Engine—Leader of Clear Processing of unmanned image".

Institutional Review Board Statement: Not applicable.

Informed Consent Statement: Not applicable.

Data Availability Statement: Not applicable.

Conflicts of Interest: The authors declare no conflict of interest. The funders had no role in the design of the study; in the collection, analysis, or interpretation of data; in the writing of the manuscript; or in the decision to publish the results.

Abbreviations

The following abbreviations are used in this manuscript:

UAV	Unmanned Aerial Vehicle
FDMPHN	Fast Deep Multi-Patch Hierarchical Network
DAMPHN	Dual Attention Level Feature Fusion Multi-Patch Hierarchical Network
PSNR	Peak Signal-to-Noise Ratio
SSIM	Structural Similarity Index Measure
APT	Average Processing Time
DCP	Dark Channel Prior
CAP	Color Decay Prior
CNN	Convolutional neural networks
AOD-Net	All-in-One Dehazing Network
DCPDN	Densely Connected Pyramid Dehazing Network
FDMSHN	Fast Deep Multi-Scale Hierarchical Network
DA	Dual Attention Level Feature Fusion
Ca_layer	Channel attention layer
Pa_layer	Pixel attention layer
PIQE	Perception-based Image Quality Evaluation

References

1. Li, X.; Li, Z.; Wang, H.; Li, W. Unmanned aerial vehicle for transmission line inspection: Status, standardization, and perspectives. *Front. Energy Res.* **2021**, *9*, 713634. [CrossRef]
2. Zhang, T.; Tang, Q.; Li, B.; Zhu, X. Genesis and dissipation mechanisms of radiation-advection fogs in Chengdu based on multiple detection data. *Meteorol. Sci. Technol.* **2019**, *47*, 70–78.
3. Zhao, L.; Zuo, X.; Zhang, S.; Lu, Y. On restoration of mountain haze image based on non-local prior algorithm. *Electron. Opt. Control* **2022**, *29*, 55–58.
4. Imran, A.; Zhu, Q.; Sulaman, M.; Bukhtiar, A.; Xu, M. Electric-Dipole Gated Two Terminal Phototransistor for Charge-Coupled Device. *Adv. Opt. Mater.* **2023**, 2300910. [CrossRef]
5. Swinehart, D.-F. The beer-lambert law. *J. Chem. Educ.* **1962**, *39*, 333–335. [CrossRef]
6. He, K.; Sun, J.; Tang, X. Single image haze removal using dark channel prior. *IEEE Trans. Pattern Anal. Mach. Intell.* **2010**, *33*, 2341–2353.
7. Zhu, Q.; Mai, J.; Shao, L. A fast single image haze removal algorithm using color attenuation prior. *IEEE Trans. Image Process.* **2015**, *24*, 3522–3533.
8. Cai, B.; Xu, X.; Jia, K.; Qing, C.; Tao, D. DehazeNet: An end-to-end system for single image haze removal. *IEEE Trans. Image Process.* **2016**, *25*, 5187–5198. [CrossRef]
9. Li, B.; Peng, X.; Wang, Z.; Xu, J.; Feng, D. AOD-Net: All-in-one dehazing network. In Proceedings of the IEEE International Conference on Computer Vision, Venice, Italy, 22–29 October 2017; pp. 4770–4778.
10. Zhang, H.; Patel, V.-M. Densely connected pyramid dehazing network. In Proceedings of the IEEE Conference on Computer Vision and Pattern Recognition, Salt Lake City, UT, USA, 18–22 June 2018; pp. 3194–3203.
11. Li, Y.; Miao, Q.; Quyang, W.; Ma, Z.; Fang, H.; Dong, C.; Quan, Y. LAP-Net: Level-aware progressive network for image dehazing. In Proceedings of the IEEE/CVF International Conference on Computer Vision, Seoul, Republic of Korea, 27 October–2 November 2019; pp. 3276–3285.
12. Li, R.; Pan, J.; He, M.; Li, Z.; Tang, J. Task-oriented network for image dehazing. *IEEE Trans. Image Process.* **2020**, *29*, 6523–6534. [CrossRef]
13. Bai, H.; Pan, J.; Xiang, X.; Tang, J. Self-guided image dehazing using progressive feature fusion. *IEEE Trans. Image Process.* **2022**, *31*, 1217–1229. [CrossRef]
14. Das, S.-D.; Dutta, S. Fast deep multi-patch hierarchical network for nonhomogeneous image dehazing. In Proceedings of the IEEE/CVF Conference on Computer Vision and Pattern Recognition Workshops, Seattle, WA, USA, 14–19 June 2020; pp. 482–483.
15. Zhang, H.; Dai, Y.; Li, H.; Koniusz, P. Deep stacked hierarchical multi-patch network for image deblurring. In Proceedings of the IEEE/CVF Conference on Computer Vision and Pattern Recognition, Long Beach, CA, USA, 15–21 June 2019; pp. 5978–5986.
16. Wang, K.; Yang, Y.; Li, B.; Cui, L. Uneven image dehazing by heterogeneous twin network. *IEEE Access* **2020**, *8*, 118485–118496. [CrossRef]
17. Liu, X.; Ma, Y.; Shi, Z.; Chen, J. Griddehazenet: Attention-based multi-scale network for image dehazing. In Proceedings of the IEEE/CVF International Conference on Computer Vision, Seoul, Republic of Korea, 27 October–2 November 2019; pp. 7314–7323.
18. Qin, X.; Wang, Z.; Bai, Y.; Xie, X.; Jia, H. FFA-Net: Feature fusion attention network for single image dehazing. In Proceedings of the AAAI Conference on Artificial Intelligence, New York, NY, USA, 7–12 February 2020; Volume 34, pp. 11908–11915.
19. Wang, C.; Shen, H.-Z.; Fan, F.; Shao, M.-W.; Yang, C.-S.; Luo, J.-C.; Deng, L.-J. EAA-Net: A novel edge assisted attention network for single image dehazing. *Knowl.-Based Syst.* **2021**, *228*, 107279. [CrossRef]

20. Yang, K.; Zhang, J.; Fang, Z. Multi-patch and multi-scale hierarchical aggregation network for fast nonhomogeneous image dehazing. *Comput. Sci.* **2021**, *48*, 250–257.
21. Wang, K.; Duan, Y.; Yang, Y.; Fei, S. Uneven hazy image dehazing based on transmitted attention mechanism. *Pattern Recognit. Artif. Intell.* **2022**, *35*, 575–588.
22. Zhao, D.; Mo, B.; Zhu, X.; Zhao, J.; Zhang, H.; Tao, Y.; Zhao, C. Dynamic Multi-Attention Dehazing Network with Adaptive Feature Fusion. *Electronics* **2023**, *12*, 529. [CrossRef]
23. Guo, Y.; Gao, Y.; Liu, W.; Lu, Y.; Qu, J.; He, S.; Ren, W. SCANet: Self-Paced Semi-Curricular Attention Network for Non-Homogeneous Image Dehazing. In Proceedings of the IEEE/CVF Conference on Computer Vision and Pattern Recognition, Vancouver, BC, Canada, 18–22 June 2023; pp. 1884–1893.
24. Liu, J.; Jia, R.; Li, W.; Ma, F.; Wang, X. Image dehazing method of transmission line for unmanned aerial vehicle inspection based on densely connection pyramid network. *Wirel. Commun. Mob. Comput.* **2020**, *2020*, 1–9.
25. Zhang, M.; Song, Z.; Yang, J.; Gao, M.; Hu, Y.; Yuan, C.; Jiang, Z.; Cheng, W. Study on the enhancement method of online monitoring image of dense fog environment with power lines in smart city. *Front. Neurorobotics* **2022**, *16*, 299. [CrossRef]
26. Zhai, Y.; Jiang, L.; Long, Y.; Zhao, Z. Dark channel prior dehazing method for transmission channel image based on sky region segmentation. *J. North China Electr. Power Univ.* **2021**, *48*, 89–97.
27. Xin, R.; Chen, X.; Wu, J.; Yang, K.; Wang, X.; Zhai, Y. Insulator Umbrella Disc Shedding Detection in Foggy Weather. *Sensors* **2023**, *22*, 4871. [CrossRef]
28. Gao, Y.; Yang, J.; Zhang, K.; Peng, H.; Wang, Y.; Xia, N.; Yao, G. A New Method of Conductor Galloping Monitoring Using the Target Detection of Infrared Source. *Electronics* **2022**, *11*, 1207. [CrossRef]
29. Yan, L.; Zai, W.; Wang, J.; Yang, D. Image Defogging Method for Transmission Channel Inspection by UAV Based on Deep Multi-patch Layered Network. In Proceedings of the Panda Forum on Power and Energy (PandaFPE), Chengdu, China, 27–30 April 2023; pp. 855–860.
30. Wang, K.; Yang, Y.; Fei, S. Review of hazy image sharpening methods. *CAAI Trans. Telligent Syst.* **2023**, *18*, 217–230.
31. Ancuti, C.-O.; Ancuti, C.; Sbert, D.; Timofte, R. Dense-haze: A benchmark for image dehazing with dense-haze and haze-free images. In Proceedings of the 2019 IEEE International Conference on Image Processing (ICIP), Taipei, Taiwan, 22–25 September 2019; pp. 1014–1018.
32. Ancuti, C.-O.; Ancuti, C.; Timofte, R.; De, C. O-haze: A dehazing benchmark with real hazy and haze-free outdoor images. In Proceedings of the IEEE Conference on Computer Vision and Pattern Recognition Workshops, Salt Lake City, UT, USA, 18–22 June 2018; pp. 754–762.
33. Ancuti, C.-O.; Ancuti, C.; Timofte, R. NH-HAZE: An image dehazing benchmark with non-homogeneous hazy and haze-free images. In Proceedings of the IEEE/CVF Conference on Computer Vision and Pattern Recognition Workshops, Seattle, WA, USA, 14–19 June 2020; pp. 444–485.
34. Zhang, N.; Zhang, L.; Cheng, Z. Towards simulating foggy and hazy images and evaluating their authenticity. In Proceedings of the Neural Information Processing: 24th International Conference, Guangzhou, China, 14–18 November 2017; pp. 405–415.
35. Harsányi, K.; Kiss, K.; Majdik, A.; Sziranyi, T. A hybrid CNN approach for single image depth estimation: A case study. In Proceedings of the International Conference on Multimedia and Network Information System, Hong Kong, China, 1 June 2019; pp. 372–381.
36. Wang, Z.; Bovik, A.-C. Mean squared error: Love it or leave it? A new look at signal fidelity measures. *IEEE Signal Process. Mag.* **2009**, *26*, 98–117. [CrossRef]
37. Wang, Z.; Bovik, A.-C.; Sheikh, H.-R.; Simoncelli, E.-P. Image quality assessment: From error visibility to structural similarity. *IEEE Trans. Image Process.* **2004**, *13*, 600–612.
38. Venkatanath, N.; Praneeth, D.; Bh, M.-C.; Channappayya, S.; Medasani, S. Blind image quality evaluation using perception based features. In Proceedings of the 2015 Twenty First National Conference on Communications (NCC), Munbai, India, 27 February–1 March 2015; pp. 1–6.
39. Amyar, A.; Modzelewski, R.; Vera, P.; Morard, V.; Ruan, S. Multi-task multi-scale learning for outcome prediction in 3D PET images. *Comput. Biol. Med.* **2022**, *151*, 106208. [CrossRef]

MDPI
St. Alban-Anlage 66
4052 Basel
Switzerland
www.mdpi.com

Sensors Editorial Office
E-mail: sensors@mdpi.com
www.mdpi.com/journal/sensors